THE BIOCHEMISTRY STUDENT COMPANION

THIRD EDITION

Allen J. Scism
Central Missouri State University

PRINCIPLES OF BIOCHEMISTRY

FOURTH EDITION

Horton   Moran   Scrimgeour   Perry   Rawn

PEARSON

Prentice
Hall

Upper Saddle River, NJ 07458

Editor-in-Chief: Dan Kaveney
Executive Editor: Gary Carlson
Project Manager: Crissy Dudonis
Executive Managing Editor: Kathleen Schiaparelli
Assistant Managing Editor: Karen Bosch
Production Editor: Diane Hernandez
Supplement Cover Management/Design: Paul Gourhan
Manufacturing Buyer: Ilene Kahn

© 2006, 2002, 1996 Pearson Education, Inc.
Pearson Prentice Hall
Pearson Education, Inc.
Upper Saddle River, NJ 07458

The author and publisher of this book have used their best efforts in preparing this book. These efforts include the development, research, and testing of the theories and programs to determine their effectiveness. The author and publisher make no warranty of any kind, expressed or implied, with regard to these programs or the documentation contained in this book. The author and publisher shall not be liable in any event for incidental or consequential damages in connection with, or arising out of, the furnishing, performance, or use of these programs.

Printed in the United States of America

10 9 8 7 6 5 4 3 2 1

ISBN 0-13-147605-X

Pearson Education Ltd., *London*
Pearson Education Australia Pty. Ltd., *Sydney*
Pearson Education Singapore, Pte. Ltd.
Pearson Education North Asia Ltd., *Hong Kong*
Pearson Education Canada, Inc., *Toronto*
Pearson Educación de Mexico, S.A. de C.V.
Pearson Education—Japan, *Tokyo*
Pearson Education Malaysia, Pte. Ltd.

Table of Contents

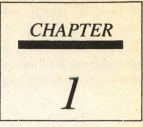
Introduction to Biochemistry

To the Biochemistry Student

A useful definition of *biochemistry* is "the tools of chemistry and physics applied to the problems of biology." You can interpret this definition in two ways. First, much of the progress in biochemistry is driven by the experimental tools of chemistry and physics, including structural analysis, kinetics, thermodynamics, radioisotope labeling, spectroscopy, crystallography, and molecular modeling. Understanding biochemistry as a dynamic, evolving field means seeing how currently accepted models have been devised by interpreting data obtained using chemical tools. Second, and of the most immediate importance as you begin your study of biochemistry, you will need to use your own tools of chemical understanding, which you developed in general and organic chemistry. This may mean that you will need to review material from previous courses, especially organic chemistry. This chapter provides a brief review of concepts from organic chemistry that you are most likely to need in biochemistry. I hope this review is adequate to get you started, but if any subject I raise here is unfamiliar, look it up in the index of your organic chemistry text to find further information.

> NOTE: Terms in *italics* are *biochemical terms*, and they may be entirely new to you, but I will explain them where they first appear. Terms in **bold** are **review terms**, and they should be somewhat familiar. Each review term will appear in one or more biochemical examples designed to help you recall the term and its meaning.

Overview

To begin this review, study Figure 1, which shows a sequence of ten chemical reactions that constitute the *metabolic pathway* called *glycolysis*. This sequence of reactions occurs in almost all living organisms. One of its main functions is to derive energy, in the form of the chemically active substance *ATP*, from the nutrient glucose. Each compound in the pathway is a product of the previous reaction and a reactant in the succeeding reaction, and each is therefore called a *metabolic intermediate*. Other reactants and products are shown entering and leaving the pathway by arrows above or below the main reaction arrows. You will study this pathway in more detail later. For now, examining it will serve to remind you of, and to illustrate, many areas of basic organic chemistry as they apply to biochemistry. For convenience, the reactions of the pathway are numbered, and each compound is labeled with its full and abbreviated name. I will use only the abbreviated names in the text. Don't worry about remembering the names and structures of all these compounds. Focus instead on the **review terms** and try to understand how they apply in these new surroundings.

Before I describe the pathway in detail, here's a sampling of the concepts I will consider. I will draw your attention first to the <u>structures</u> of the intermediates, and second to the <u>reactions</u> depicted. First, look at structures. Try giving **systematic names** to a few of them, using *oxyphosphoryl* as the **substituent** name of the OPO_3^{2-} group. Your names will not be the same as the commonly used names; for example, the oxyphosphoryl group is usually called a **phosphate** group, as in glucose 6-phosphate, the product of the first reaction. Among the intermediates, can you find pairs of **structural isomers**? What **functional groups** can you find and name? What functional groups are stabilized by **resonance**? Do you remember the meaning of resonance, and how to draw the most important structures that contribute to a **resonance hybrid**? Which intermediates are **chiral compounds** and how many **stereoisomers** are possible for each structure? Biochemists use D– and L– to designate the **configurations** of chiral compounds. Recall that each **chiral**

center can also be assigned the (**R**)- or (**S**)-configuration, according to the system of Cahn, Ingold, and Prelog. Which molecules might exist in equilibrium with forms other than those shown? For example, which might exist in two or more **tautomeric forms**, and which contain two functional groups that might combine with each other in an **intramolecular reaction** to produce an alternative form?

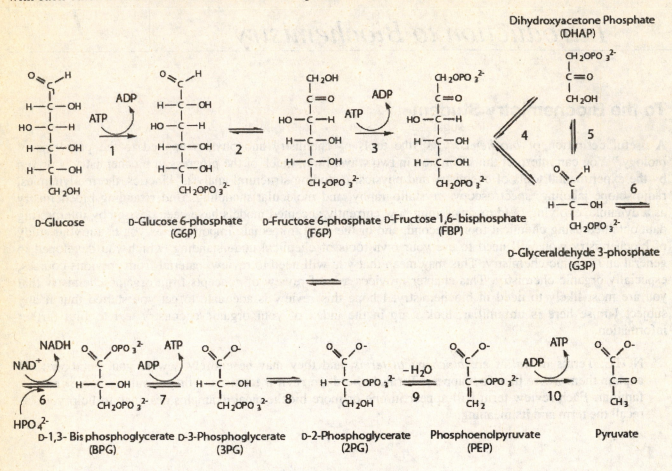

Figure 1. The Ten Reactions of Glycolysis

Second, examine each reaction. Try to discern what structural change converts the reactant to the product, and try to classify each change as **addition**, **elimination**, **substitution**, or combinations of these simple **reaction types**. Can you write **mechanisms** for these reactions, assuming some simple type of **catalysis**, like **specific-acid catalysis** (catalysis by protons), and depicting with **curved arrows** the changing allegiances of electrons as bonds break and form? (This powerful form of writing reaction mechanisms is sometimes called **electron pushing**.) Do you see any **oxidation** or **reduction** reactions? One way to detect such changes is to compute the average **oxidation number** of carbon for each structure, assuming that the oxidation number of hydrogen is +1 and that of oxygen is -2. (To simplify detection of oxidation-number changes in carbon, you can ignore complex groups that do not change during the reaction.)

Now you have an idea of some of the areas of organic chemistry that are essential to understanding a metabolic pathway, and some of the concepts that you need to review and use in your study of biochemistry. Let's begin our detailed review by looking at the pathway more closely.

The Structures

The first structure in Figure 1 could be called 2,3,4,5,6-pentahydroxyhexanal, but fortunately it has a common name, **D-glucose**. In fact, many biological compounds have common names that were adopted before their structures were known. This prevalence of common or trivial names is in part a blessing, because the names are often much shorter than systematic names, and in part a curse, because you cannot figure out

2

the names from the structures. You must therefore simply memorize many new names. Looking at the structure of glucose, you can probably deduce that this compound is water-soluble. Organic compounds having at least one **hydroxyl** (–OH) or other **hydrogen-bonding** group per three or four carbons are usually appreciably soluble in water. Looking over all the intermediates in glycolysis, you can see that all would be very soluble in water. Not all biomolecules are water-soluble, however. *Fats and oils*, which are **esters** of *fatty acids* such as *palmitic acid* ($CH_3(CH_2)_{14}COOH$), are quite insoluble in water because they are mostly **nonpolar**. They are quite soluble in organic solvents like hexane. Many water-insoluble biomolecules belong to the class called *lipids*. Some are involved in forming *membranes*, which provide a barrier that makes the cell and its internal compartments impermeable to many water-soluble substances.

The sugar glucose is a ***monosaccharide***, a simple ***carbohydrate***. Carbohydrates, one of the four main types or classes of biomolecules, include sugars and starches (the other biomolecules are *lipids*, *proteins*, and *nucleic acids*). Most monosaccharides are polyhydroxycarbonyl compounds, either **aldehydes** like glucose, or **ketones**. Notice that glucose has four **stereocenters** or **chiral centers**, at carbons 2, 3, 4, and 5 (the aldehyde carbon is C-1). These carbons form covalent bonds (which means that they share electron pairs) through sp^3-**hybridized orbitals**, so they form four bonds with **tetrahedral** geometry. (C-1, in contrast, is sp^2-**hybridized** and **trigonal planar**.) Because two **configurations** are possible at each stereocenter, there are 2^4 or 16 different **stereoisomers** of 2,3,4,5,6-pentahydroxyhexanal. Only the full mirror image, or the **enantiomer**, of D-glucose, shares its name. The enantiomer shown is D-glucose, designated D- because of its structural kinship with the simpler chiral compound D-*glyceraldehyde* (the 3-phosphate derivative of D-glyceraldehyde, called D-glyceraldehye 3-phosphate, is one of the products of Reaction 4 in Figure 1). The enantiomer of D-glucose is called L-glucose. The other stereoisomers of glucose, which are its **diastereoisomers** or **diastereomers**, are chemically distinct from D- and L-glucose, and have their own common names.

The drawings of glucose and the other intermediates in Figure 1 are called **Fischer projections**. At each chiral center, the groups attached by horizontal bonds in Fischer projections are understood to be projected outward toward the reader, while the groups attached by vertical bonds are pointing back into the page. Because the chiral carbons are sp^3 hybridized, the angle formed by the chiral carbon and any pair of attached atoms is approximately 109.5 degrees.

Because glucose is an aldehyde with a hydrogen atom on its α-carbon (C-2), in aqueous solutions it exists in equilibrium with a **tautomeric enol** form. The equilibrium reaction is shown in Figure 2.

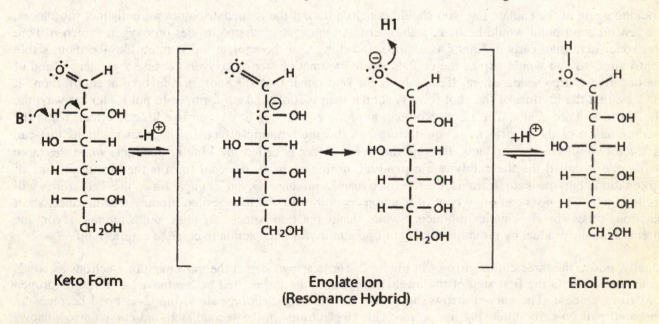

Keto Form   **Enolate Ion** (Resonance Hybrid)   **Enol Form**

Figure 2. Tautomerism in Glucose

Carbons next to carbonyls are weakly acidic. If a base abstracts a proton from the α-carbon of the aldehyde (which is called the **keto** tautomer), a resonance-stabilized **enolate ion** (shown in square brackets) results. The enolate may be reprotonated (almost certainly by a different proton from the solvent) at C-2 or at O-1 (the oxygen of C-1). Protonation at O-1 (shown) produces the enol form of glucose. This process is reversible, and because the reactant, glucose, is less acidic than the enol, it is more stable and predominates in the tautomeric equilibrium. The fraction of enol-glucose at equilibrium is so small that we always write glucose as shown in Figure 1, but tautomerism is important in explaining both spectroscopic and chemical properties of aldehydes and ketones, as I will show later.

Now examine the structures shown within the square brackets in Figure 2. These structures, taken together, depict the enolate ion as a **resonance hybrid** of the structures shown, which are called **resonance contributors**. Organic chemists depict molecules or ions as resonance hybrids when one Lewis diagram is inadequate to depict them adequately. In this enolate ion, the negative charge is distributed between C-2 and the oxygen of C-1, so neither of the two drawings alone fully depicts the properties of the ion. This terminology does <u>not</u> imply that an enolate is an equilibrium mixture of the resonance contributors. It is better to say that the enolate ion is a <u>composite</u> of the contributors, which means that it has some of the properties of both.

A telling analogy from an organic text of the late sixties helped me understand the meaning of resonance. It goes like this:

> A rhinoceros is a resonance hybrid of a dragon and a unicorn. This does not imply that the rhino spends part of its time as a dragon and part of its time as a unicorn. Instead, it means that the rhino has some dragon attributes (like a tough, thick coat), and some unicorn attributes (like its horn). In addition the rhino, like the resonance hybrid, is a real animal, but the dragon and unicorn, like the contributors, are fictional.

So actually, the resonance contributors do not exist at all. They are fictional characters whose structures suggest the properties we want to depict for the real enolate ion. We view the hybrid as being most like the contributor or contributors we judge to be the most stable, based on criteria such as charge distribution and number of covalent bonds. Contributors with less separation of opposite charge and more shared electrons are more stable. And the more equivalent forms we can write, the greater is the stabilization by resonance. (While I think of the contributors as useful fictions, I leave for philosophers the question of whether molecules and ions themselves are real. Like the vivid characters in good fiction, they certainly help us to see the real world more clearly.)

Looking again at the enolate ion, you should conclude that, if the two contributors were distinct substances, the second contributor would be more stable because the negative charge resides on oxygen, which is more electronegative than carbon. Chemists assume, therefore, that the enolate ion is more like the more stable contributor, so you would expect the enolate ion to protonate faster on oxygen because a greater portion of the negative charge resides there. If so, why is the keto rather than the enol form favored at equilibrium? It must be that the lifetime of the enol form is shorter than that of the keto form, or to put it another way, the enol form is more acidic. The faster protonation of the enolate ion at oxygen is a **kinetic** effect, while the predominance of the keto form at equilibrium is a **thermodynamic** effect. In chemical reactions that can produce two alternative products, the one that forms faster is called the **kinetic product**, while the more stable one is called the **thermodynamic product**. In this example, the enol form is the kinetic product of protonation, but the keto form is the thermodynamic product. Given enough time, the keto form will predominate. Chemists can often control reaction conditions to obtain a desired product. For instance, short reaction times favor kinetic products, while long reaction times at high temperatures favor the thermodynamic product by providing plenty of opportunity for the reaction to come to equilibrium.

Finally, notice the three curved arrows in Figure 2. These arrows depict the movement of electrons as bonds break and form. In the first step of the mechanism shown, an unspecified base removes a proton (hydrogen ion) from glucose. The curved arrow shows that the bonding electron pair in the C–H bond becomes an unshared pair on carbon, leaving the proton with no electrons. In the second step, the curved arrow shows that an unshared electron pair on oxygen becomes a bonding pair between oxygen and a proton. Notice that the arrows depict the movement of *electrons*, not nuclei -- for example, no arrow follows the lost proton. In writing reaction mechanisms, curved arrows are used only to depict electron movement.

Now focus on the various functional groups of the intermediates in Figure 1. Glucose, with its **carbonyl** (–C=O) group at C-1, is an **aldehyde**. It also possesses **hydroxyl** (–OH) groups, the functional groups found in **alcohols**. The product of Reaction 2 in Figure 1, fructose 6-phosphate, or F6P, is a **ketone**, because the carbonyl carbon is attached to two carbons. Like glucose, it is also an alcohol. Joined to O-6 of fructose is a **phosphate** ($-OPO_3^{2-}$) group. F6P is thus also an **ester**, which is the product of reaction between an alcohol and an acid. You are most familiar with esters of **carboxylic acids**, which contain **carboxyl** groups (-COOH). F6P is the product of reaction between an alcohol (the C-6 hydroxyl of fructose) and phosphoric acid. Figure 3a compares a carboxylate ester with a phosphate ester. Notice the similarity in bonding in the ester linkages (in boxes).

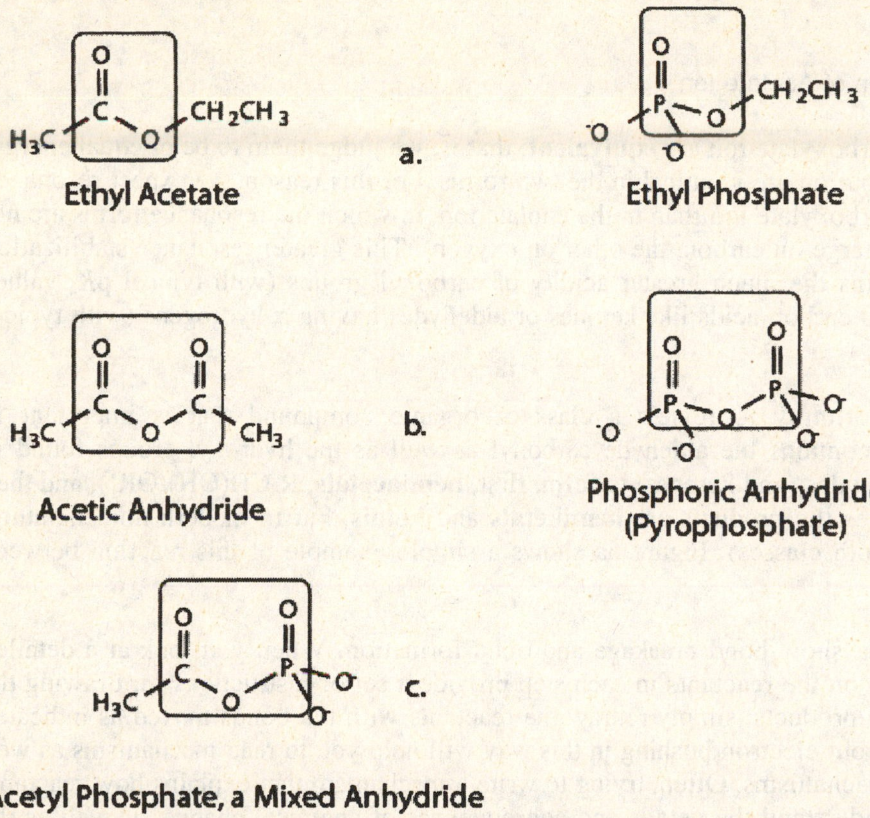

Ethyl Acetate

Ethyl Phosphate

a.

Acetic Anhydride

Phosphoric Anhydride
(Pyrophosphate)

b.

Acetyl Phosphate, a Mixed Anhydride

c.

Figure 3. Analogous Carbon and Phosphorus Compounds **a.** Carboxylate and Phosphate Esters **b.** Carboxylate and Phosphate Anhydrides **c.** A Mixed Anhydride

The product of Reaction 6 (Figure 1), BPG, is an **anhydride**. As with esters, you are most familiar with anhydrides of carboxylic acids. 1,3-*Bis*phosphoglyceric acid (BPG) is a **mixed anhydride** of phosphoric acid and a carboxylic acid, 3-phosphoglyceric acid. Figures 3b and 3c compare carboxylic, mixed carboxylic-phosphoric, and phosphoric anhydrides (called **phosphoanhydrides**) Notice the similarity in bonding in the anhydride linkages (circled). **Hydrolysis** (literally, cleavage by water) of anhydrides gives acids.

Reaction 7 produces a carboxylic acid (3PG) by transfer of the phosphate from BPG to *adenosine 5'-diphosphate* (*ADP*). The carboxyl group of 3PG is shown in Figure 1 in its unprotonated, or **carboxylate**, form, because this form would predominate at pH values common to the cellular and intracellular fluids of living organisms (pH 7 to 8.5). The pK_a values for simple carboxylic acids are well below 7. When the pH of the solution is numerically the same as the pK_a of an acid, the acid is 50% ionized; that is, the protonated and unprotonated forms are present in equal quantities. At higher pH (lower concentration of hydrogen ions), the unprotonated form predominates by a factor of 10 for each unit of difference between the pH and the pK_a. For example, if the pK_a of an acid is 4.0, then at pH 7.0, the unprotonated groups (carboxylates) outnumber the protonated groups (carboxyls) by 1000 to 1. At pH 4.0, carboxylates and carboxyls are present in equal quantities.

The carboxylate group provides another example of resonance stabilization, as shown for acetate ion in Figure 4. Curved arrows on the first contributor show that you can use electron pushing to help you write resonance contributors. See if you can use curved arrows to generate the second contributor for the enolate ion in Figure 2.

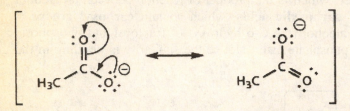

Figure 4. Resonance Stabilization of Acetate Ion

The two resonance forms of the carboxylate ion are **equivalent**; that is, we judge them to be of equal energy, because bonding and charge distribution are identical in the two forms. For this reason, we expect resonance stabilization to be greater in the carboxylate ion than in the enolate ion, in which the resonance forms are not equivalent (one carries negative charge on carbon, the other on oxygen). This greater resonance stabilization of the carboxylate partially explains the much greater acidity of carboxyl groups (with typical pK_a values between 2 and 6) in comparison to carbon acids like ketones or aldehydes having α-hydrogens (with typical pK_a values greater than 20).

Glucose provides one more opportunity to review a class of organic compound that is important in carbohydrate chemistry. Glucose contains the aldehyde carbonyl as well as the hydroxyl groups found in alcohols. Alcohols react with aldehydes (and ketones) to form, first, **hemiacetals**, R-CH(OH)(OR'), and then **acetals**, R-CH(OR')$_2$, (with ketones the products are **hemiketals** and **ketals**, but in modern nomenclature, *acetal* is the preferred term for both classes). Figure 5a shows a simple example of this reaction between benzaldehyde and methanol.

Again, note the curved arrows that show bond breakage and bond formation. When you look at a detailed mechanism, notice that the arrows on the reactants in each step provide a set of instructions for drawing the products of that step. To draw the products, simply redraw the reactants with the bonds moved as indicated by the curved arrows. Thinking about electron pushing in this way will help you to read mechanisms as well as to write reasonable and clear mechanisms. Often, trying to write a mechanism that explains how reactants becomes products helps you to understand the details and consequences of chemical change, as well as the roles of the various agents (like hydrogen ions) that promote reactions.

The presence of both the aldehyde carbonyl and hydroxyl groups in glucose makes possible intramolecular hemiacetal formation. Intramolecular reactions are particularly likely when they produce five- or six-membered rings. It is therefore not surprising that glucose in water forms a cyclic hemiacetal. The first step in Figure 5b shows the formation of this cyclic form, called glucopyranose, from the open-chain form of glucose. The hydroxyl of C-5 reacts with the C-1 carbonyl of form a six-membered ring. See if you can write a mechanism for this process, using the mechanism in Figure 5a as a guide. Assume acid catalysis.

When you encounter glycolysis later in your course, you will probably find that glucose is drawn in this glucopyranose form, which predominates in solution. The hemiacetal hydroxyl, on C-1 in the case of glucopyranose, is quite reactive in substitution reactions with other alcohols or amines. Other sugar molecules can join with glucose by way of their hydroxyls to form **glucosides** such as complex sugars and starches, which are acetals of glucose. The general term for an acetal of a monosaccharide is **glycoside**. The second step in Figure 5b shows the formation of a methyl glycoside from glucose. Some sugars, like ribose, react with amines to form *N*-glycosides, as shown in Figure 5c You will find examples of *N*-glycosides in DNA and RNA. Try writing a mechanisms for the reaction in Figure 5c. Assume acid catalysis.

Amines and **thiols** are among other compounds that are important in biochemistry. Amines, whether **primary** (RNH$_2$), **secondary** (R$_2$NH), or **tertiary** (R$_3$N), are organic bases. The pK_a values for the conjugate acids of most amines (for instance, RNH$_3^+$, a **primary ammonium ion**) are higher than 9.0, so protonated forms of amines predominate under physiological conditions. Simple **thiols** (RSH) have pK_a values around

9.0, so a significant amount of the thiolate form (RS⁻) can exist at neutral pH, although RSH predominates. Both thiols and the thiolates are powerful **nucleophiles**. An important reaction of thiols is oxidation to **disulfides** (RSSR). This reaction occurs spontaneously in the presence of the oxygen in air. Disulfides stabilize the structures of some proteins.

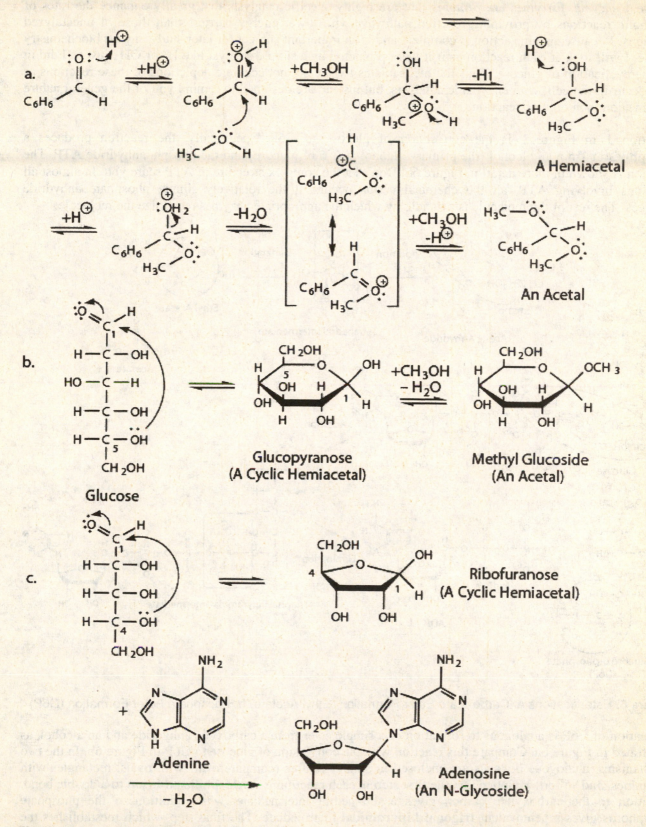

Figure 5. Hemiacetals and Acetals **a.** Reaction of Benzaldehyde with Methanol **b.** Formation of the Cyclic Hemiacetal of Glucose and Methyl Glucoside, an Acetal **c.** Formation of the Cyclic Hemiacetal of Ribose and Adenosine, an N-glycoside

The Reactions

Having used some of the intermediates of glycolysis to review many aspects of structural organic chemistry, let's go through the pathway step by step, looking at reactions. In the cell, each reaction is catalyzed by a different *enzyme*. Enzymes are *proteins* that are highly specific catalysts that greatly enhance the rates of metabolic reactions by providing reaction pathways with lower energy barriers than those of uncatalyzed reactions. Because enzyme action is complex, and is an important subject for later study in your biochemistry course, I will look at each reaction as if it were promoted by a simple catalyst like H^+ or OH^-, just as I did in the the cyclization of glucose. Thus the mechanisms I write are not accurate depictions of these reactions as they occur in the cell, with enzymatic catalysis, but instead are designed to remind you of the general nature of some common organic reactions.

Reaction 1 in Figure 1 is the conversion of glucose to G6P. Specifically, the reaction produces a **phosphoester** from an alcohol (the primary alcohol at C-6 of glucose) and the phosphoanhydride ATP. The reaction is shown in more detail in Figure 6b. Don't let the complex structure of ATP scare you. In almost all reactions involving ATP, all the chemical action occurs at the relatively simple phosphate anhydride linkages. The rest of the molecule is a handle by which the appropriate enzymes recognize the molecule.

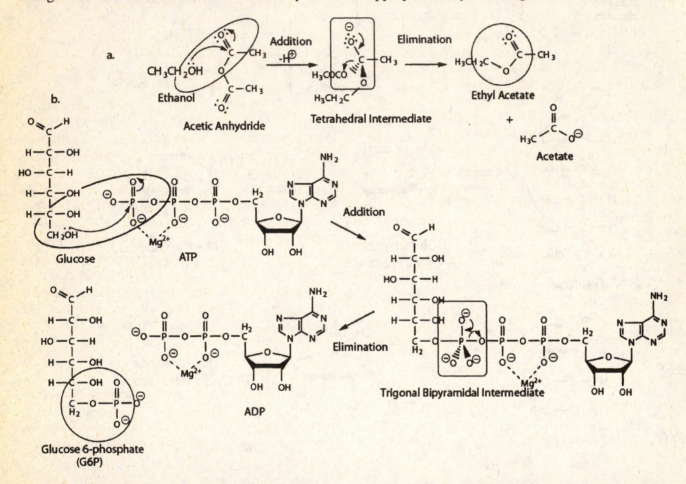

Figure 6. Esterifications **a.** Carboxylate Ester Formation (Ethyl Acetate) **b.** Phosphate Ester Formation (G6P)

Formation of G6P is analogous to formation of a simple ester from a carboxylic anhydride and an alcohol, as illustrated in Figure 6a. Compare this reaction with the conversion of glucose to G6P in Figure 6b. In the two mechanisms, analogous linkages are enclosed in similar figures (compare ovals with ovals, rectangles with rectangles, and so forth). Notice that the first step in each reaction is **nucleophilic addition** to a double bond. Addition to the carboxylate carbon gives a tetrahedral intermediate, while addition to the phosphate phosphorus gives a pentavalent **trigonal bipyramidal** intermediate. The final step, which reestablishes the double bond, is **elimination**. The two steps together constitute a **substitution** reaction. Derivatives of carboxylic and phosphoric acids readily undergo this type of substitution.

The reactions of Figure 6 occur spontaneously under standard conditions (that is, their equilibrium constants are greater than 1.0) because anhydrides like ATP are thermodynamically less stable than esters like G6P. Thus the reactants (anhydride plus alcohol) are less stable than the products (ester plus acid), and products predominate at equilibrium. In general, anhydrides are the least stable compounds derived from carboxylic acids. Roughly speaking, the order of thermodynamic reactivity for biologically important carboxyl derivatives and analogous phosphate compounds is as follows:

$$\text{carboxyl anhydride} \geq \text{mixed (carboxyl-phosphate) anhydride} > \text{phosphoanhydride} \geq \text{thioester (RCOSR')} > \text{ester} > \text{amide (RCONH}_2\text{)} > \text{acid}$$

This order largely reflects the stability of the group that leave the carbonyl carbon during hydrolysis. The most reactive carboxyl derivatives release the least basic leaving groups. Further, this order reflects thermodynamic stability, which is related to equilibrium constants for reactions, as opposed to kinetic stability, which is related to rates of reactions. You can use this order of reactivity to decide whether a reaction has a favorable equilibrium constant ($K > 1$). In general, the formation of any compound type from one before it in the reactivity series is favorable. For example, because esters are higher in the series than amides, treatment of an ester with an amine to form an amide is a favorable process. Referring again to Reaction 1, the formation of the phosphate ester G6P from the phosphanhydride ATP is a favorable process.

All reactions involving ATP require the presence of a divalent metal ion, most commonly Mg^{2+}. Mg^{2+} forms complexes with phosphate oxygen atoms on ATP. Several different types of complexes are possible, and representative structures are shown in Figure 6b. Mg^{2+} partially neutralizes negative charges on the oxygen atoms, and thus makes ATP more susceptible to nucleophilic attack. As a result, Mg^{2+}—ATP is a better **electrophile** than free ATP. This role is one of many for metal ions in biochemistry. Ions are essential components of some enzymes, where they may act as **Lewis acids** (electron acceptors) or as **oxidation-reduction** (**redox**) agents. Some biochemistry texts represent bonding to metal ions with dotted lines, implying electrostatic attraction, while others represent it with solid lines, implying covalent bonding. Inorganic chemists use several bonding theories to explain bonding in metal-ion complexes, but it is usually safe to interpret interactions between metal ions and electron donating species, called **ligands**, as being electrostatic in nature, with ligands donating electrons into empty orbitals of the metal ion. You can judge the hybridization of the metal ion by the geometry of the complex, the most common cases being **octahedral** (d^2sp^3, 6 ligands around the metal ion, or a **coordination number** of 6), **trigonal bipyramidal** (dsp^3, 5 ligands), and **tetrahedral** (sp^3, 4 ligands).

Reaction 2, the formation of F6P from G6P, produces a ketone from an aldehyde. Look again at the enol form of glucose in Figure 2. You can see that it is in fact an **enediol**, because both carbons of the double bond carry hydroxyls. This means that a third tautomer is possible, with a carbonyl at C-2. This tautomer is fructose. It is thought that the conversion of G6P to F6P proceeds by way of this enediol intermediate. Try to write a four-step mechanism for this process (the three intermediates are 1. enolate ion; 2. enediol; 3. enolate ion). Assume acid catalysis.

In Reaction 3 of Figure 1, F6P accepts a second phosphate group from ATP, producing FBP, in a reaction similar to Reaction 1. In Reaction 4, FBP is split into two 3-carbon sugars by a reaction whose reverse is a familiar carbon-carbon bond-forming reaction, the **aldol condensation**. Figure 7 compares a simple reverse aldol reaction (a) with reaction 4 (b). In Figure 7a, the reverse-aldol cleavage of a β-hydroxy aldehyde is catalyzed by hydroxide ion, an example of **specific base catalysis**. In Figure 7b, the reaction is catalyzed by an unspecified base, which could be a basic functional group on the enzyme that catalyzes this reaction. This is an example of **general base catalysis**, which simply means catalysis by a base other than hydroxide ion.

A reverse aldol reaction requires a β-hydroxy aldehyde or ketone. Because most monosaccharides contain hydroxyl groups at their β positions (two carbons away from the carbonyl, or at C-3 in FBP), reverse aldol reactions are common in carbohydrate chemistry.

Of the two products from reaction 4, only G3P can take part in Reaction 6. DHAP is not wasted, however, because it is converted to a second molecule of G3P by Reaction 5, which is very similar to Reaction 2. Write a mechanism for the acid-catalyzed formation of G3P from DHAP, by analogy to the mechanism of Reaction 2. Both molecules of G3P continue through the pathway, so two molecules of pyruvate are formed from each molecule of glucose that enters glycolysis. This conclusion can be tested by **radioisotope labeling** experiments, in which the fate of specific atoms can be followed by replacing them with radioactive isotopes in a small fraction of the molecules. For example, if FBP containing the radioactive isotope ^{14}C at position C-4 is added to a cell extract undergoing active glycolysis, all of the radioactivity is recovered in C-1 of pyruvate. If the experiment is repeated using ^{14}C-3 FBP, the radioactivity is again found in C-1 of pyruvate, demonstrating that both DHAP and G3P are converted to pyruvate by the enzymes of glycolysis. Biochemists have used radioisotopic labeling extensively in establishing the mechanisms of reactions and the order of intermediates in metabolic pathways.

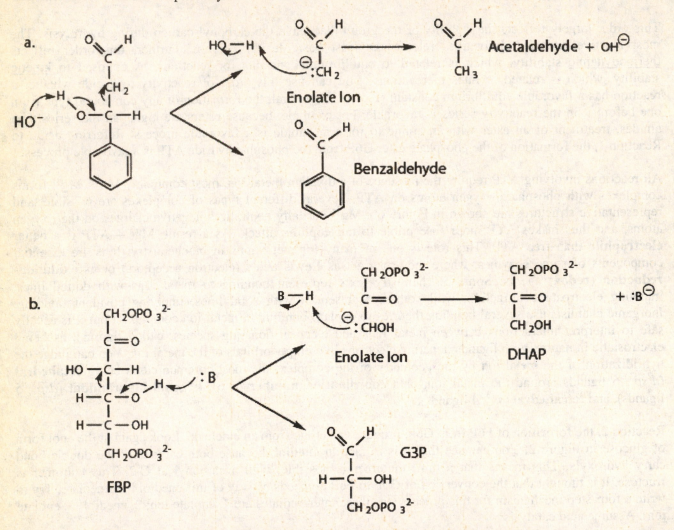

Figure 7. a. Reverse Aldol Reaction **b.** Conversion of FBP to DHAP and G3P

Reaction 6 converts G3P and a phosphate ion to 1,3-BPG. To help recognize what type of reaction this is, calculate the **oxidation number** of C-1 in the reactant and product. Here is a two-step procedure for computing oxidation numbers within a complete structural formula. First, assign all electrons to atoms as follows: assign unshared electrons to the atom on which they reside, and bonding electrons to the more electronegative of the two bonded atoms; if a bond joins two atoms of the same element, assign one bonding electron to each. Second, compute the oxidation number for any atom by subtracting the number of assigned electrons from the group number of that element in the Periodic Table. For C-1 of G3P, the oxidation number is [(group number for carbon) − (assigned electrons)] = 4 − 3 = +1. The oxidation number of C-1 in 1,3-BPG is 4 − 2 = +2. Reaction 6 thus increases the oxidation number of C-1, which means that this reaction is an

oxidation.

Oxidation is always accompanied by **reduction**. The substance that becomes reduced during the reaction is called the **oxidizing agent**. The oxidizing agent in Reaction 6 is NAD^+, and the reduced product is NADH. NAD^+ is derived from nicotinamide, a *vitamin*. The oxidized and reduced forms of the nicotinamide ring, which are the reactive parts of the structures of NAD^+ and NADH, are shown in Figure 8 (the R group, like most of the ATP molecule, is a large, complex "handle" for enzymes). C-4 of NAD^+ is thought to accept a hydride ion (H^-) from C-1 of G3P in this reaction. As an exercise in computing oxidation numbers, calculate oxidation numbers for C-4 in NAD^+ and in NADH, and thus convince yourself that this carbon is reduced (its oxidation number <u>decreases</u>) during Reaction 6.

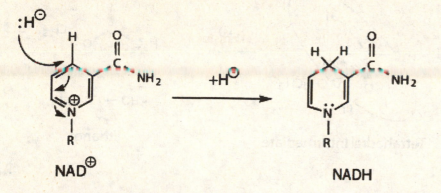

Figure 8. Reduction of NAD^+ to NADH

Among biological oxidizing agents, NAD^+ is relatively weak. For example the equilibrium constant for an NAD^+-dependent oxidation is about 10^{38} times smaller than that for the same oxidation by O_2, one of the cell's most powerful oxidizing agents. It takes a powerful reducing agent like the aldehyde of G3P to reduce NAD^+ to NADH. Of course, the reduced form of a weak oxidizer is a strong reducer, and reduced nicotinamides like NADH act as powerful reducing agents in the synthesis of many biomolecules. The reducing power of NADH is due at least in part to the fact that when NADH acts as a reducing agent, the product, NAD^+, is **aromatic**, and hence particularly stable. Notice in Figure 8 that the nicotinamide ring of NAD^+ contains six atoms and three alternating double bonds. You can therefore write two equivalent resonance contributors for this ring, suggesting that it is significantly stabilized by resonance, and suggesting further that the *p*-orbitals of the six sp^2-hybridized atoms of the ring form a continuous, cyclic π system containing six delocalized electrons. Quantum mechanical models suggest that special stability attends such cyclic systems when the number of delocalized electrons can be expressed as $4n + 2$, in which *n* is any whole number. Compounds with this electronic configuration (2, or 6, or 10, ... π electrons) are called **aromatic**. The most familiar aromatic compound is benzene (C_6H_6), but aromatics can also contain **heteroatoms** (non-carbon atoms), such as the nitrogen in the **heteroaromatic** ring of NAD^+.

In Reaction 7, the phosphate of 1,3-BPG is transferred to ADP to form ATP. The mechanism is similar to the reverse of Reactions 1 and 3, but note that 1,3-BPG is a mixed anhydride (carboxylate-phosphate) and not a phosphoester like G6P or FBP. Because mixed anhydrides are more reactive than phosphoanhydrides, transfer of phosphate from 1,3-BPG to ADP is favorable. The reaction is reversible, however, and serves as a step in the synthesis of glucose, as well as its degradation. The product of the forward reaction, ATP, is a source of chemical energy that is used to drive many cellular processes.

Reaction 8, the conversion of 3PG to 2PG, appears to be simply transfer of the phosphate group from one hydroxyl (at C-3) to another (at C-2). This is an example of **transesterification**. Figure 9a shows a typical organic transesterification, in which the acyl group of an ester is transferred to an alcohol, leaving a different alcohol behind. See if you can fill in the mechanistic details of this reaction, using a hydrogen ion as the catalyst. The first step is protonation of the carbonyl oxygen.

Figure 9b shows a plausible mechanism for Reaction 8. Analogous intermediates in Figures 9a and 9b are circled. When you study glycolysis, you will learn that this reaction is more complex than it appears here, and that the phosphate is not simply transferred directly from C-3 to C-2. As evidence of a more complex mechanism, in yeast and muscle, each molecule of the enzyme that catalyzes this reaction is known to contain one phosphate group. If the phosphate groups of the enzymes are labeled with the radioisotope ^{32}P, the radioactive phosphorus appears in the phosphate of 2PG, suggesting that the enzyme swaps the phosphate group of 3PG for its own phosphate group. In fact, a dephosphorylated, inactive enzyme can be restored to full activity by a stoichiometric amount of 2,3-*bis*-phosphoglycerate (2,3-BPG, shown in Figure 9c), suggesting that 2,3-BPG is an intermediate in the reaction. This is another example of the power of radioisotopic labeling in revealing details of reaction mechanisms.

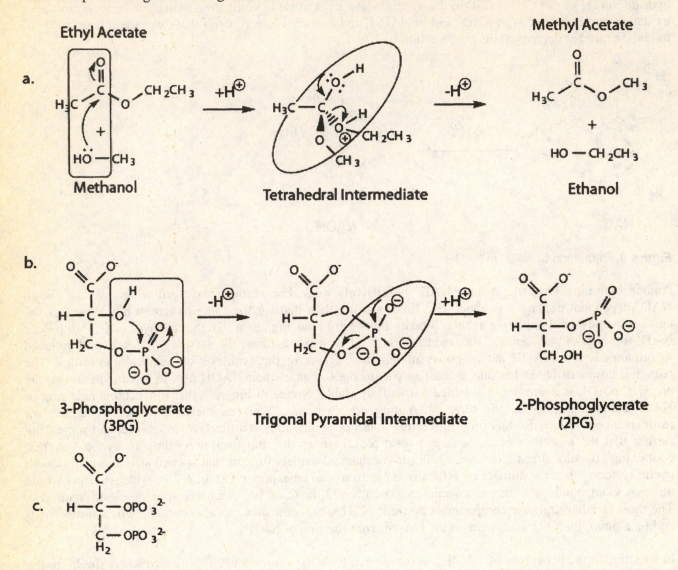

Figure 9. Transesterification **a.** Conversion of Ethyl Acetate to Methyl Acetate **b.** Conversion of 3PG to 2PG: A Plausible, But Incorrect Mechanism **c.** 2,3 DPG

Reaction 9 is an example of **elimination**. The elements of one water molecule, a hydrogen from C-2 and the OH from C-3, are eliminated from 2PG to produce PEP. Figure 10 provides a mechanism, as if catalyzed by H$^+$. Again notice the use of curved arrows in this mechanism. In step 1, the curved arrow shows protonation of O-3, with an unshared electron pair of the oxygen becoming a bonding pair in the OH$_2^+$ group. In step 2, the hydrogen atom of C-2 leaves, and its bonding electron pair becomes a second bonding pair, making a double bond between C-2 and C-3. This new bond displaces a water molecule, which is a much better leaving group than was the C-3 OH before it was protonated. In this mechanism, notice that one proton is taken up (step 1) and one is released (step 2), so this process would not alter the concentration of hydrogen ions. Thus we can say that H$^+$ is truly a **catalyst** in this process, because its concentration is not changed by the overall reaction.

In Reaction 10, another phosphate transfer occurs, from O-2 of PEP to ADP to form ATP and pyruvate. This process looks at first glance like conversion of a phosphate ester (PEP) to an anhydride, which you would not expect to be favorable according to the reactivity series. The reason that this reaction is favorable is that the initial product of the phosphate transfer is not pyruvate, but enolpyruvate, which quickly tautomerizes to pyruvate. In this keto-enol equilibrium, pyruvate like most keto forms, is greatly favored. Remember that when one or more equilibria occur in sequence, the overall equilibrium constant is the product of the equilibrium constants for the individual steps. The large equilibrium constant for conversion of enolpyruvate to pyruvate makes the overall equilibrium favorable for Reaction 10, although the phosphate transfer itself may be much less favorable. Another way to picture this is to note that the direct product of the phosphate transfer, enolpyruvate, is removed immediately by tautomerism, which drives the reaction to completion. **Le Chatelier's principle**, which states that equilibria spontaneously shift in the direction that minimizes changes in concentrations of reactants or products, captures the logic of this view. Recall that many chemical reactions go to completion because products, for instance gases or precipitates, are lost from solution when they form.

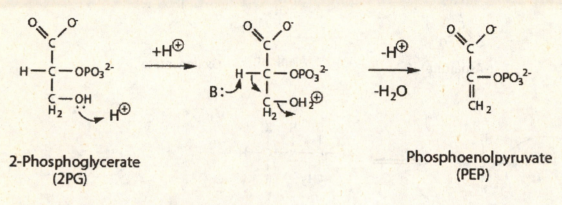

2-Phosphoglycerate
(2PG)

Phosphoenolpyruvate
(PEP)

Figure 10. Acid-Catalyzed Elimination of Water

In a living organism, what happens to the pyruvate that results from glycolysis? Like all the compounds in Figure 1, pyruvate is a metabolic intermediate. It occupies a branch point in metabolism, and can enter several other pathways. Some cells break down pyruvate further, ultimately to CO_2, in oxidative processes that require the participation of oxygen from the air. Others, like muscle cells during heavy exercise, reduce pyruvate to lactate ($CH_3CHOHCOO^-$). The overall reaction

$$C_6H_6O_6 \rightarrow 2\ C_3H_5O_3^- + 2H^+$$

acidifies muscle and causes cramps. Still others cells, like yeasts, convert pyruvate to ethanol (CH_3CH_2OH) and CO_2. This two-step process begins with the **decarboxylation** of pyruvate to form acetaldehyde (CH_3CHO), which is then reduced to ethanol. The loss of CO_2 drives this process to completion. Both lactate and ethanol production entail **reduction** by NADH, and this reduction exactly balances the oxidation in Reaction 6. The conversion of glucose to lactate or to ethanol and CO_2 therefore entails no net oxidation or reduction. (To convince yourself of this, calculate the average oxidation number for carbon in glucose and in lactate.) Because such redox-balanced processes can be sustained in the absence of oxygen, they are referred to as *anaerobic*. Indeed, anaerobic (oxygen-free) conditions are necessary for production of ethanol by yeasts during manufacture of beer and wine. If air is present, yeasts oxidize glucose to CO_2 and produce no ethanol.

Pyruvate can also be converted to the **α-amino acid** alanine (Figure 11a), one of the building blocks required for the synthesis of proteins.

One of the paths from pyruvate to alanine provides another important example in this review, the reaction of amines with ketones (or aldehydes) to form **imines (Schiff bases)**. Pyruvate, an α-ketoacid, receives its amine group in a **transamination** reaction with an α-amino acid, which in turn becomes an α-ketoacid. Figure 11a shows this process, with the amino acid aspartate as the source of the amino group, and oxaloacetate as the by-product. Figure 11b shows with general structures how a ketone and an amine can exchange an amine group. First, they join to form an imine (Figure 11b, imine A) in a **condensation** (water-forming) reaction. This process is reversible, and the reverse reaction, hydrolysis of imine A, would regenerate the reactants. But imine A equilibrates (another example of tautomerism) with imine B, which is also subject to hydrolysis. Hydrolysis of imine B completes the transfer of the amino group from one reactant to the other. Try writing a mechanism for the hydrolysis of imine B, by analogy with the reverse of the initial condensation that forms imine A. Biological transaminations are more complex than shown in Figure 11, involving a third amine carrier that receives the amine group from an α-amino acid and then transfers it to an α-ketoacid. Both transfers entail a mechanism similar to that shown in Figure 11b. Imines that are subject to tautomeric shifts that move the double bond from one side of the nitrogen to the other are instrumental in a number of group transfers like transamination.

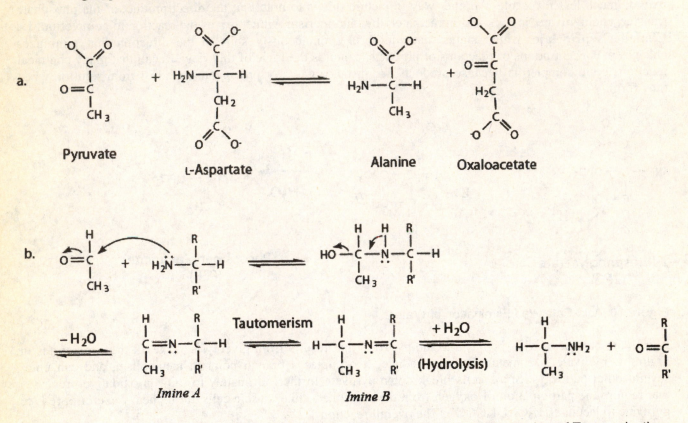

Figure 11. Transaminations **a.** Formation of Alanine from Pyruvate **b.** General Mechanism of Transamination

As mentioned earlier, α-amino acids like alanine are the building blocks of proteins. Proteins are **polymers** constructed from α-amino acids that are joined into long chains by condensation reactions. The process is complex, but the overall reaction is the condensation of the α-amino group of one amino acid with the α-carboxyl group of the next, so that the two are joined by an **amide** linkage, which in proteins is called a *peptide bond*. When two amino acids are joined by a peptide bond (Figure 12a), the product is called a *dipeptide*. Water is a by product. There are twenty different α-amino acids that the cell can use in protein synthesis, and proteins range in size from less than 100 to more than 1000 amino-acid residues. Figure 12b shows a portion of a protein chain, with the *residues* (the remains after loss of water) of two amino acids marked by parentheses. The potential number of different proteins containing 100 amino acids is 20^{100} (or about 10^{130}), since each of the 100 positions in the chain can be any one of 20 different amino acids. The universe contains far fewer than 10^{130} atoms (about 10^{80}). It is no wonder that cells can make a bewildering array of proteins from a small number of building blocks. Each enzyme in glycolysis, as well as each enzyme in every other metabolic pathway, is a protein. Even relatively simple bacterial cells contain thousands of different proteins.

Enzymes Versus Chemical Catalysts

Let me remind you that all the reactions in glycolysis are catalyzed by enzymes (each of which is a protein), and that the mechanisms of the reactions are more complex than I have shown here. In these examples, I have simply written *plausible* mechanisms to provide a review of basic skills like electron pushing. Remember that a chemist proposes a mechanism to explain all available experimental evidence concerning a reaction. Realistic mechanisms, like the ones you find in your biochemistry text, are acceptable only if they satisfy a number of requirements. For instance, they must specify a role for all reagents known to be essential to the reaction, including specific parts of the enzyme; they must include all intermediates that have been detected in the reaction, for example, by spectroscopy or chemical trapping; and they must be compatible with results of isotopic labeling experiments, with the stereochemical consequences of the reaction, and with measurements of reaction **kinetics** (dependence of reaction rates on reaction conditions). In short, a mechanism is an attempt to explain the course of a reaction in a way that accounts for all that is known about the process. The quest to determine the mechanisms of biological reactions pushes mechanistic organic chemistry to its greatest heights.

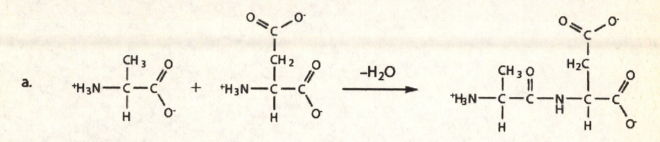

Two α-Amino Acids

A Dipeptide

A Portion of a Protein

Figure 12. Amides **a.** Condensation of Two α-Amino Acids to Form a Dipeptide **b.** Proteins are Polymers of α-Amino Acids

Enzymes are far more powerful and selective agents than chemical agents like hydrogen ions. As an example, consider Reaction 1 again. Any of the five hydroxyl groups of D-glucose could just as readily accept the phosphate group from ATP; all five hydroxyls have about the same chemical properties. But the enzyme that catalyzes this reaction <u>directs</u> the new phosphate group to C-6 only, despite the almost identical reactivity of the other hydroxyls. This is an example of **regiospecificity** in enzymatic catalysis: enzymes can distinguish chemically identical but structurally distinct functional groups, and thus enzymes do not produce the side products that plague laboratory reactions.

As a further example, notice that Reaction 4, like most of the reactions in glycolysis, is a reversible reaction. This means that in the reverse reaction, the enzyme assembles FBP from DHAP and G3P. The process in this direction produces two new chiral centers: C-3 and C-4 of FBP. In this enzyme-catalyzed aldol condensation, only one of four possible diastereoisomers is produced: FBP itself. Any <u>chemically</u> catalyzed version of this reaction would produce all four products (the chiral center of G3P would result in unequal amounts of the four diastereomers, an example of **asymmetric induction**). The enzyme that catalyzes Reaction 4 splits only FBP in the forward direction, and produces only FBP in the reverse direction. This is an example of **stereospecificity** in enzymatic catalysis: enzymes can distinguish stereoisomers, even enantiomers, as reactants, and can produce specific stereoisomers as products, again without the side products that are inevitable in most laboratory reactions. No wonder that biochemists are seeking ways to use enzymes as commercial catalysts.

An understanding, at the molecular level, of the power of enzymes is just one of many interesting insights that await you in your study of biochemistry. You are embarking on a study of one of the most powerful, fast-moving, and exciting areas of science. I hope that this review of important concepts from organic chemistry helps you to get off on the right foot in your biochemistry course.

Water

I. KEY TERMS

With the help of this study guide, your textbook, and class notes, you should be able to define and explain the significance of the following terms:

acid
amphipathic
anions
base
buffered
cations
chaotrope
charge-charge interaction
conjugate acid
conjugate base
denatured

electrolytes
electrophile
Henderson-Hasselbalch
equation
hydrated
hydrogen bond
hydrophilic
hydrophobic
hydrophobic effect
hydrophobic interaction
ion pairing

micelle
nucleophile
osmosis
osmotic pressure
pH
pK_a
polar
solvated
solvation sphere
van der Waals forces

II. EXERCISES

$pH = pka + log \dfrac{[A^-]}{[HA]}$

A. True-False

_____F____ **1.** The O–H bonds in water are polar due to the high electronegativity of hydrogen.

_____T____ **2.** Water is a polar molecule (has a dipole moment) because it is V-shaped.

_____T____ **3.** In the liquid state, each water molecule has the potential to form hydrogen bonds with four other water molecules.

＊ _____T____ **4.** Compounds that dissolve readily in water are termed hydrophilic and include electrolytes as well as nonelectrolytes.

＊? _____F____ **5.** The osmotic pressure of an aqueous 0.001 M starch solution is greater than that of an aqueous 0.001 M glucose solution. *depends on [ ]!*

＊ _____T____ **6.** Polar molecules can induce polarity in nonpolar molecules.

_____T____ **7.** An aqueous solution of pH 3.0 contains H_2O, H^+, and OH^-.

_____F____ **8.** An acid that dissociates to the extent of 88% in water would be termed a strong acid.

? _____T____ **9.** A solution that contains 0.05 mol of lactic acid and 0.05 mol of potassium lactate per liter of solution has a pH of 3.86.

$pH = pka + log \dfrac{[A^-]}{[HA]}$

_____T____ **10.** The major buffer system of the blood is the bicarbonate-carbonic acid buffer system.

? _____F____ **11.** The shell of water molecules that surrounds a dissolved ion or molecule is called a hydrosphere. *solvation sphere*

_____T_____ 12. Although they operate over similar distances, hydrogen bonds are much stronger than van der Waals forces.

_____T_____ 13. In a buffer system, increasing the concentration of the conjugate base relative to that of the conjugate acid increases pH. _(but slightly at first)_

B. Short Answers

1. _electronegativity_ is the tendency of an atom to attract to itself the shared electrons in a covalent bond.

✱ 2. _Chaotropic_ agents are certain ions and molecules that are poorly solvated in water and enhance the solubility of nonpolar compounds by disordering the water molecules.

3. _Amphipathic_ molecules contain both hydrophobic and hydrophilic groups.

4. Soap molecules dissolved in water tend to form aggregates called _micelle_.

— 5. A shell of water molecules that forms around ions as they become dissolved is called a/an _solvation sphere_

6. Attractive forces between molecules, collectively called _Van der Waals_, can occur due to dipole-dipole attractions, ion-dipole attractions, and interactions between two nonpolar atoms or molecules that result in transiently induced dipoles, which are known as London dispersion forces.

7. The unit(s) of K_w, the ion-product constant of water, is(are) _M^2_. $K_w = [OH^-][H^+]$

8. An acid or base that dissociates less than 100% in water is described as being _weak_.

9. The unit(s) of K_a, the dissociation constant for a weak acid, is(are) _M (mol/L)_
$K_e = \frac{[A^-][H^+]}{[HA]}$

10. A solution that contains equal or nearly equal quantities of a weak acid and its conjugate base is called a/an _buffer_.

11. The condition called _acidosis_ results when a person's blood pH becomes lower than normal.

12. Compounds that dissolve in water to produce ions that can conduct a current through the resulting solution are called _electrophytes_.

⊙ 13. The strongest buffer in blood after the carbon dioxide-carbonic acid-carbonate system is _hemoglobin_

C. Problems

1. The electron-pair geometry in water is tetrahedral. Why, then, is the angle between the two O–H bonds 104.5° rather than 109.5°?

✗ 2. Inside cells, what is the most important factor that results in slowing the rate of diffusion of molecules? _molecular crowding_

3. Calculate the pH of (a) a solution of 0.0100 M acetic acid and (b) a solution of 0.0100 M HCl. How do they compare? $CH_3COOH \rightleftharpoons CH_3COO^- + H^+$ .01 M, 0, 0 .01-x, x, x $Ka = \frac{x^2}{.01-x}$ $\to x = 4.2 \times 10^{-4}$ $pH = -\log[.01] = 2.00$

4. In a solution of 0.00350 M HNO₃ (SA), evaluate [H⁺], pH, and [OH⁻]. $pH = -\log[H^+]$ $pH = -\log[.0035]$ $[H^+] = .00350$ $Kw = [H^+][OH^-] = 1 \times 10^{-14}$

5. In an acetic acid-acetate buffer of pH 4.25, what is the ratio of acetate to acetic acid? $4.85 = pKa + \log\frac{[A^-]}{[HA]}$

6. Your text indicates that the energy required to break each hydrogen bond in ice is 23 kJ/mol (section 2.2) and that four hydrogen bonds are possible for each H_2O molecule. Assuming this bond energy is the same for ice and liquid water at 0.0°C, but that 78.2 kJ/mol is the actual energy required to break the hydrogen bonds in the liquid water, determine the average number of hydrogen bonds per molecule that exist in the liquid water in contrast with the four per molecule in ice. Express your answer as a number such as 1.7, 2.6 or 3.9, etc., bonds per water molecule. $\frac{78.2}{23}$ 3.5

7. Write the formula of the species that, together with H_2O, constitutes a conjugate acid-base pair. What does your answer say about the properties of H_2O? amphoteric (can be both acid + base)

$H_2O + H_2O \rightleftharpoons H_3O^+ + OH^-$

8. Here is a thought question some instructors have used: What is the pH of a 10^{-8} HCl solution?

$[H^+] = 10^{-8}$ + $[H^+]$ in H_2O at this low dilution

9. What is the pH of 0.0105 M lactic acid? $pH = pka + log \frac{[A^-]}{[HA]}$ weak acid $Ka = \frac{[H^+]^2}{[HA]}$ for weak acids at low concentration

* 10. During the titration of 50.0 mL of 0.0500 M lactic acid with 0.0500 M NaOH, the biochemist stopped after 25.0 mL of the base had been added. Can this solution be used as a buffer? What is the pH of the solution?

Half of la. have had their pt removed by NaOH   yes

11. What is the pH of a solution containing 0.0350 M benzoic acid and 0.0400 M sodium benzoate? The pK_a for benzoic acid is 3.19. $pH = 3.19 + log \frac{[.04]}{[.035]}$

12. At the normal physiological pH of 7.4, the ratio of bicarbonate to carbonic acid in the blood is about 10.7:1. This would not appear to be an ideal buffer since the two buffer components are not in approximately equal quantities. How is this system able to adequately neutralize OH^-?

HCO_3^- H_2CO_3

$H_2O + CO_2 \rightleftharpoons H_2CO_3$ As H_2CO_3 is used, more is formed from CO_2 carried in blood.

D. Additional Problems

1. Calculate the pH of a solution made by adding 50.0 mL of 0.025 M NH_3 to 50.0 mL of 0.030 M NH_4Cl.

$pH = pka + log \frac{[NH_4^+]}{[NH_3]} = 9.2 + log$

2. Refer to Problem C.3. in which you contrasted the pH of 0.0100 M HCl with that of 0.0100 M acetic acid. From that information and the solution to that problem, calculate the % ionization of acetic acid in that solution.

3. If the blood pH in a human became 7.28, what would be the ratio of HCO_3^- to H_2CO_3 in the blood?

4. At 25°C, water ionizes such that the value of K_w is 1.00×10^{-14}. Water ionizes more extensively at higher temperatures. If $K_w = 1.4 \times 10^{-14}$ at some temperature greater than 25°C, what is the pH at this elevated temperature? Is the system acidic, basic, or neutral?   pH scale is based on particular temp!

5. What is the pH of a solution prepared by mixing 600.0 mL of 0.075 M KH_2PO_4 with 400.0 mL of 0.12 M K_2HPO_4? $[H_2PO_4^-] = .6 \times .075 = .045$ $[HPO_4^{2-}] = .4 \times .12 = .048$ $pH = pka + log \frac{[.048]}{[.045]}$

6. If the pH of distilled water in equilibrium with air is about 5.68, what is the maximum possible concentration of dissolved carbon dioxide? Use information in text Table 2.4.

$pH (H_2O) = 5.68$ $H_2O \rightleftharpoons H_2O$ $pH = pka + $

7. Calculate the grams of K_2HPO_4 one would need to dissolve in 1.00 L of 0.0800 M KH_2PO_4 to create a buffer of pH 7.30.

$7.3 = 7.2 + log \frac{[HPO_4^{2-}]}{[H_2PO_4^-]}$ $[HPO_4^{2-}] = [H_2PO_4^-]$ $.08 mol \frac{1 L}{1 mol}$ why is this term gone?

$CO_2 + H_2O \rightleftharpoons H_2CO_3 \rightleftharpoons 2H^+ + CO_3^{2-}$ $Ka = \frac{[H^+]^2}{[H_2CO_3]}$

$5.68 = -log[H^+]$ $[H^+] = 2 \times 10^{-6}$

x   →0   0
x-y   2y   y

Amino Acids and the
Primary Structures of Proteins

I. KEY TERMS

With the help of this study guide, your textbook, and class notes, you should be able to define and explain the significance of the following terms:

affinity chromatography	eluate	microenvironment
amino acid analysis	enantiomers	N-terminus
chiral	fractionation	peptide bond
column chromatography	gel-filtration chromatography	polyacrylamide gel
configuration	HPLC	electrophoresis
C-terminus	hydropathy	primary structure
disulfide bridge	ion-exchange chromatography	sequenator
Edman degradation procedure	isoelectric point	stereoisomers
electrophoresis	MALDI	zwitterion
electrospray mass spectrometry	mass spectrometry	

II. EXERCISES

A. True-False

_____T_____ 1.   In a medium of pH 2.0, aspartic acid has a net positive charge.

_____T_____ 2.   There is no pH condition at which the predominant form of lysine would be a species with three separate charges.

_____F_____ 3.   The majority of the amino acids used to make <u>natural proteins</u> have highly hydrophilic R-groups.

_____T_____ 4.   One mole of aspartic acid in its fully protonated form would require three moles of OH⁻ to convert it to the fully unprotonated form.

_____F_____ 5.   All 20 of the amino acids used to make natural proteins are optically active.

_____X T_____ 6.   The isoelectric point for most of the amino acids that have nonpolar side chains is about six. *where side gp*
    $pI = \frac{(pKa_1 + pKa_2)}{2}$ (for others, $pI = \frac{(pK_1/2) + pKa}{2}$) *Depends* *pKa falls on curve*

_____F_____ 7.   Since the common amino acids have the L– stereo configuration, one may assume that they will rotate plane-polarized light in a left-handed (levorotatory) direction.
    Depends on side group.

_____F_____ 8.   The addition of 0.015 mole of L-valine to a solution containing 0.015 mole of D-valine would produce a solution that would exhibit twice the degree of rotation of polarized light that the D-valine solution had before the addition.

_____F_____ 9.   A linear molecule made from amino acids linked end to end, which has four peptide bonds in its structure, is called a tetrapeptide.
    pentapeptide

_____F_____ 10. All of the α-L-amino acids with only one chiral carbon have the (S) designation in the *RS* system.

_F___ **11.** When electrophoresis is performed on a mixture of methionine and aspartic acid buffered at pH 5.7, methionine will move toward the cathode and aspartic acid will move toward the anode.

Ⓧ ~~F~~ T **12.** In SDS-PAGE, proteins are separated from each other primarily on the basis of their size and not their charge. _Both_ → all are neg-charged, so only size matters

Ⓧ _F_ **13.** Complete amino acid analysis of a peptide is possible using acid hydrolysis with, for example, 6 M HCl at 110°C for 16–72 hours. Ⓧ destroys some!

_T_ **14.** It is easier to determine the sequence of a protein by sequencing the DNA of the gene that codes for that protein than to purify and directly sequence the protein itself.

B. Short Answers

1. In partition column chromatography, one or more solutes may be retained on the column while another passes through with little or no retention. When a solute is removed from the column, it is said to have been ___eluted___.

2. An amino acid in a dipolar ion form can be called a/an ___zwitterion___.

3. The pH at which a given amino acid carries a net zero charge is referred to as ___isoelectric pt___.

4. The covalent linkage formed by the oxidation of the side chains of two cysteine residues in a peptide or protein is called a/an ___disulf. bond___

5. A measure of the relative hydrophobicity or hydrophilicity of a given amino acid is called its _____.

6. The immediate surroundings of a side chain of a given amino acid residue in a protein are called its _____.

7. D-leucine and L-leucine may be called a pair of ___enantiomers___

8. The specific sequence of amino acid residues in a peptide or protein is called its ___1° structure___

9. The _____ can be used to determine the sequence of a peptide by means of an automated instrument called a/an _____.

10. The sequence of a protein or peptide is given direction by indicating its primary structure from its ___N-term___ to its ___C-term___.

11. ___Arg [R]___ is the most basic of the 20 common amino acids.

12. Polyacrylamide gel electrophoresis, PAGE, separates proteins on the basis of ___charge___ and ___mass___.

C. Problems

1. Of 100 molecules of alanine, how many have ionized carboxyl groups at pH 3.0?

pka [COO⁻] = 2.0

$3 = 2.0 + \log \dfrac{[COO^-]}{[COOH]}$

$4 = \dfrac{x}{100-x}$

$400 - 4x = x$

$400 = 5x$

$80 = x$

2. Stereoisomers that are not enantiomers are called diastereomers. Draw a diastereomer of the form of α-L-threonine shown here.

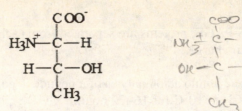

$$
\begin{array}{c}
COO^- \\
H_3N^+\!-\!C\!-\!H \\
H\!-\!C\!-\!OH \\
CH_3
\end{array}
$$

3. Why is glycine a crystalline solid rather than a liquid as are many other organic molecules of similar molecular weight? *Doesn't rotate light*

4. Draw the complete, charged structure of glycylvalylalanine at pH 7.

5. Draw the ionic forms of serine that predominate at each pH listed below.
   (a) 2.0 (b) 5.7 (c) 10.0   *a) NH₃ ⊕ COOH   c) NH₂ ⊕ COO⁻*
   b) NH₃ ⊕ COO⁻

6. Explain the electrophoretic migration behavior you would expect of aspartame at (a) pH 3.0 and (b) pH 6.9. Assume that the pKa values of the free amino acids given in text Table 3.2 are applicable for this dipeptide. *a) COO⁻, NH₃, COOH only ¹/100, so toward cathode*
   b) COO⁻, NH₃, COO⁻   pI = (3.9+9.9)/2 = 6.9 neutral! wouldn't migrate
   2   9.9   3.9

7. Draw an enantiomer and a diastereomer of this molecule.

$$
\begin{array}{c}
CHO \\
H\!-\!C\!-\!OH \\
H_3N^+\!-\!C\!-\!H \\
CH_2OH
\end{array}
$$

8. Calculate the isoelectric points of
   (a) valine, (b) lysine, and (c) aspartic acid.

9. Use graph paper to draw the titration curve (pH on the y-axis vs. equivalents of OH⁻ on the x-axis) for one mole of an amino acid with pK_{a1} = 2.4 and pK_{a2} = 9.6. Label the following points on the curve.
   (a) pK_{a1} (α-carboxyl group)
   (b) pK_{a2} (α-amino group)
   (c) pI (isoelectric point)

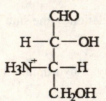

10. Assign the appropriate *RS* designation to this molecule: 2-amino-4-bromo-4-hydroxybutanoic acid.

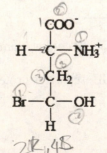

$$
\begin{array}{c}
COO^- \\
H\!-\!C\!-\!NH_3^+ \\
CH_2 \\
Br\!-\!C\!-\!OH \\
H
\end{array}
$$

2R, 4S

11. With respect to cellular protein synthesis, what is significant about the sequence of a protein chain as mentioned in question B.10 above?

12. In SDS-PAGE, all proteins bind SDS such that they all have approximately the same charge-to-mass ratios. How, then, is separation achieved?

   by size (smaller molec. move faster).

D. Additional Problems

1. Calculate the ratio of conjugate base to acid form for each of the ionizable groups of glutamic acid in a medium of pH 4.00.

2. Using the letters A, V, and L, show the primary structure of all possible peptides that contain one residue each of L-alanine, L-valine, and L-leucine.

3. How many different peptides of 15 residues can be made from the 20 common amino acids?

4. You have a mixture of L-lysine, L-aspartate, and glycine that is subjected to electrophoresis in a buffered solution of pH 6. How will the amino acids behave? Explain.

5. The labels have been lost from the containers of two dipeptide samples. One is known to contain a residue each of alanine and glycine. The other contains a residue each of alanine and aspartic acid. A 1.0 mmole sample of one of the dipeptides in its fully protonated form required 30. mL of 0.10 M NaOH to convert it to its totally unprotonated form. Which dipeptide is it?

6. Using the hexapeptide Gly–Met–Arg–Asp–Phe–Gly, show the results of treating separate samples of this peptide with (a) trypsin, (b) BrCN, (c) *S. aureus* V8, and (d) chymotrypsin.

7. A small peptide was found to contain equimolar amounts of the following amino acids: arginine, glutamic acid, glycine, lysine, methionine, and phenylalanine. Individual samples of the peptide were treated with the following agents with the results noted:
   (a) trypsin: L-arginine and a pentapeptide.
   (b) cyanogen bromide: two tripeptides.
   (c) *S. aureus* V8: L-lysine and a pentapeptide.
   (d) chymotrypsin: a dipeptide and a tetrapeptide, with the latter showing absorbance at 260 nm.

   What is the primary structure of the peptide? Explain each piece of evidence given and the reasoning that led to your answer.

8. Explain how SDS-PAGE and mass spectrometry can be used in a complementary manner to gain rather accurate information about a protein.

Proteins: Three-Dimensional Structure and Function

I. KEY TERMS

With the help of this study guide, your textbook, and class notes, you should be able to define and explain the significance of the following terms:

2,3-*bis*phosphoglycerate
3_{10} helix
allosteric effector
allosteric interaction
allosteric modulator
allosteric protein
antibody
antigen
Bohr effect
capping box
carbamate adduct
conformation
cooperativity of folding
denaturation
domain
fibrous protein
fold
globular protein

hairpin loop
heat shock proteins
hypervariable region
immunoassay
immunoglobulin fold
loop
molecular chaperone
monomer
motif
multienzyme complex
native conformation
oligomer
oxygenation
peptide group
pitch
positive cooperativity of
    binding
primary structure

proteomics
quaternary structure
R state
random coil
rise
Schiff base
secondary structure
T state
tertiary structure
turns
α-helix
β-pleated sheet
β-sheet
β-strand
β-turns

II. EXERCISES

A. True-False

_____ 1. Secondary structural features of a protein are stabilized by covalent bonds.

_____ 2. Nearly all peptide groups in proteins studied to date have been found to have the *trans* conformation.

_____ 3. The tertiary structure of a globular protein is its overall three-dimensional shape.

_____ 4. Metabolic proteins such as enzymes are globular proteins.

_____ 5. Myoglobin binds O_2 more readily than does hemoglobin.

_____ 6. Fetal hemoglobin has a lower P_{50} than does adult hemoglobin, which indicates that fetal hemoglobin binds O_2 less strongly than does adult hemoglobin.

_____ 7. Oxygen-binding characteristics of hemoglobin proved to be quite different from those of myoglobin due to radical differences in the primary structure of their respective protein chains.

_____ 8. The rise per amino acid residue in a segment of α-helix is called the pitch.

_____ 9. The heme group of hemoglobin binds oxygen more strongly alone than it does when present in the hemoglobin molecule.

_____ 10. Recognizable combinations of α-helices and β-strands that appear in a number of different proteins are called domains.

_____ 11. The hydrophilic side chains of amino acid residues normally locate themselves toward the exterior of globular proteins.

_____ 12. Denaturation destroys the biological activity of a protein by breaking many covalent bonds in the protein.

_____ 13. The quaternary structure of proteins is maintained predominately by noncovalent interactions.

_____ 14. Hemoglobin binds oxygen more strongly as the pH is lowered.

B. Short Answers

1. _____ is the name of a new area of study that involves large sets of proteins.

2. The decreased affinity of hemoglobin for oxygen due to elevated levels of carbon dioxide and H^+ is called _____.

3. The type of secondary structure involving 75% of the residues in myoglobin is _____.

4. Two modes of denaturation that primarily affect salt bridges and disrupt hydrogen bonding, respectively, in a protein are_____ and _____.

5. The form of the protein chain that occurs upon denaturation is called _____.

6. Proteins that possess quaternary structure are called _____.

7. The actual site of oxygen binding in the heme of myoglobin is _____.

8. A/an _____ is the term for an oligomeric protein responsible for catalyzing several different metabolic reactions.

9. The process by which a protein in a random coil conformation assumes its native shape is called _____.

10. The sigmoidal curve for the binding of oxygen by hemoglobin illustrates the phenomenon of _____.

11. An organic molecule in erythrocytes that lowers the affinity of hemoglobin for oxygen is _____.

12. _____are proteins that assist in the cellular process of protein folding such that a large fraction of the protein molecules affected achieve their native conformation and are thus biologically active.

13. A region in a globular protein characterized by negative Φ and Ψ values is a region of _____.

14. In the immune response, foreign compounds called _____ cause the synthesis of proteins called _____ that combine with and precipitate the foreign compounds and mark them for destruction.

15. The activity of some proteins, such as hemoglobin, is modulated by specific small molecules called _____. Such proteins are called _____.

C. Problems

1. A segment of a protein chain in the α-helical conformation contains 20 residues. How long is this segment?

2. How many turns are in this helical segment of 20 amino acids?

3. Explain the manner in which hydrogen bonding occurs in a region of α-helix.

4. Consider the hydrogen bonding in an α-helical segment of 10 residues. Considering the potential for hydrogen bond formation, what percentage of these potential hydrogen bonds actually exists?

5. Consider the following segment of a protein chain and speculate on the length(s) of α-helix that is/are feasible.

$$-Asp-Met-Phe-Pro-Ala-Met-Ala-His-Leu-Ile-Pro-$$

6. Scurvy is a human disease characterized by skin lesions, loose teeth, and bruises on arms and legs due to fragile blood vessels. What is the connection between scurvy, vitamin C, and collagen?

7. Do covalent crosslinks form between collagen molecules? Explain your answer.

8. The enzyme RNase A is denatured, and rendered inactive, by treatment with 8 M urea containing 2-mercaptoethanol. If both urea and the 2-mercaptoethanol are removed by dialysis and the denatured protein is exposed to air, biological activity is regained. Alternatively, if the denatured protein is reoxidized in the presence of urea, only about 1% of the original activity is regained. What conclusion(s) might be made from these observations? Explain.

9. Relatively small changes in pH have rather large effects on the binding of oxygen by hemoglobin. If the pO_2 is 30 torr in certain tissues, is more, or less oxygen bound at pH 7.6 compared with that bound at pH 7.2? Express the amount of oxygen bound at pH 7.6 versus that at pH 7.2 as a ratio.

10. Your text indicates that some carbon dioxide is carried by hemoglobin due to the formation of carbamate adducts with $-NH_2$ groups of hemoglobin. Most CO_2 is transported as dissolved bicarbonate ions.
   (a) Write the reaction for the formation of carbamate adducts.
   (b) Indicate how dissolved bicarbonate ions are involved in the eventual release of CO_2 at the lungs.
   (c) Speculate on the importance of the carbamate adduct formation to transport CO_2 in terms of the overall amount of CO_2 that must be transported.

11. What is an immunoassay? Explain the underlying basis for this technique that allows it to be used, in some cases, for diagnostic tests.

12. It has been determined that globular proteins crystallize in their *in vivo* native conformations. Explain the evidence that supports this statement.

13. What is a leucine zipper?

D. Additional Problems

1. The approximate molar mass of an oligomeric protein was found by gel filtration to be 8×10^4. The native protein was shown to be composed of a single type of protein chain (monomer), the molar mass of which was determined by SDS-PAGE to be 1.9×10^4. Speculate on the oligomeric composition of the native protein.

2. The energy required to break a carbon-carbon bond is about 345 kJ/mol (Holtzclaw, *General Chemistry*, 9th Edition, 1991, page 194). Hydrogen bonds in proteins help stabilize secondary and tertiary structure. How many hydrogen bonds are required to provide the stabilization that would be the energy equivalent of one carbon-carbon bond?

3. The enzyme trypsin contains six disulfide bonds. In one study, renaturation of the denatured enzyme, following removal of the denaturant, yielded only 8% of its original activity. (a) What would be the expected random reformation of the six disulfide bridges? (b) Why is nearly full activity not regained as it was with RNase A?

4. Calculate the stabilization energy due to hydrogen bonding in the myoglobin molecule.

5. What is the percent of extensibility of keratin in hair?

6. What is the percent of deoxymyoglobin (Mb) in a solution in which the partial pressure of oxygen, pO_2, is 6.0 torr? The P_{50} for myoglobin is about 2.8 torr.

7. Estimates are that the free energy required to denature a native protein is only about 0.4 kJ/mol of amino acid residues. In a myoglobin molecule, how many hydrogen bonds need be disrupted to cause denaturation?

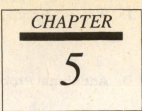

Properties of Enzymes

I. KEY TERMS

With the help of this study guide, your textbook, and class notes, you should be able to define and explain the significance of the following terms:

active site	initial velocity	rate equation
affinity label	irreversible inhibitor	reaction specificity
allosteric modulator	isomerase	regulated enzyme
catalyst	kinetic mechanism	regulatory site
catalytic constant	kinetic order	reversible inhibitor
catalytic proficiency	ligase	saturation
competitive inhibition	Lineweaver-Burk plot	second order reaction
concerted theory	lyase	sequential reaction
coupled reactions	maximum velocity, V_{max}	sequential theory
covalent modification	metabolite channeling	specificity constant, k_{cat}/K_m
dehydrogenase	Michaelis constant, K_m	stereospecificity
enzymatic reaction	Michaelis-Menten equation	substrate
enzyme	noncompetitive inhibition	synthase
enzyme assay	ordered mechanism	synthetase
enzyme-substrate complex	oxidoreductase	transferase
first order reaction	ping-pong reaction	turnover number
hydrolase	product	uncompetitive inhibition
inhibition constant	random mechanism	velocity
inhibitor	rate acceleration	zero order reaction

II. EXERCISES

A. True-False

_____ 1.   The only biological catalysts are the proteins called enzymes.

_____ 2.   All enzymes require cofactors such as FAD and coenzyme A.

_____ 3.   An enzyme that catalyzes the addition of a phosphoryl group to glucose would probably not catalyze the same reaction for glycerol.

_____ 4.   Any form of contact between an enzyme and its substrate will lead to reaction.

_____ 5.   The formation of an enzyme-substrate complex is an unsupported hypothesis.

_____ 6.   A multisubstrate, enzyme-catalyzed reaction in which a product is released before all substrates have been bound to the enzyme is an example of a sequential kinetic mechanism.

_____ 7. K_m and V_{max} values can best be determined graphically from a plot of initial velocity, v_o, versus substrate concentration, [S]. This type of plot is known as a Michaelis-Menten plot.

_____ 8. The higher the K_m value, the greater the affinity of an enzyme for its substrate.

_____ 9. The K_m values of enzymes for their substrates are usually slightly higher than the intracellular concentrations of their substrates.

_____ 10. A plot of kinetic data, v_o *vs.* [S], that produces a sigmoidal curve indicates that cooperative interactions occur between enzyme subunits.

_____ 11. Regulatory enzymes are oligomeric proteins that catalyze reactions at a committed step in a pathway.

_____ 12. Irreversible enzyme inhibitors form covalent bonds with side chains of the enzyme active site residues.

B. Short Answers

1. The region of an enzyme molecule with which the substrate must interact in order for catalysis to occur is called the _____.

2. The short-lived species formed when enzyme and substrate interact initially is the _____.

3. Enzymes that catalyze the conversion of a molecule into its structural isomer would belong to the IUBMB category of _____.

4. The number of catalytic events catalyzed per second per enzyme molecule (or per active site) is called the _____.

5. The ratio k_{cat}/K_m is called the _____ and is a measure of the _____.

6. Some enzymes are subject to control of activity by the addition and removal of phosphate groups. This is called regulation by _____.

7. Due to its structure, a/an _____ inhibitor binds in the active site of an enzyme.

8. An inhibitor that does not alter the K_m of the enzyme is a/an _____ inhibitor.

9. An inhibitor that alters both the K_m and V_{max} of the enzyme system is a/an _____ inhibitor.

10. Inhibition of a regulatory enzyme by an end product of the pathway is called _____ inhibition.

11. Sites where allosteric modulators bind to enzymes are called _____.

12. An allosteric modulator can be either a/an _____ or a/an _____.

13. An accessory enzyme that catalyzes the covalent substitution of an interconvertible enzyme, causing it to change from one conformation to another, is called a/an _____.

C. Problems

1. What are the four main properties of enzymes, according to your text?

2. What are coupled reactions? How do they benefit a biological system?

3. Sketch a graph of rate (velocity) vs. [E] for an enzyme catalyzed system that contains saturating levels of substrate. Explain other necessary conditions for such a determination.

4. The enzyme alcohol dehydrogenase catalyzes the interconversion of ethanol and acetaldehyde and the interconversion of NAD^+ and NADH. Start with acetaldehyde, the enzyme, and the appropriate form of the coenzyme and write the equation for this reaction using the structures of the substrate, product, and the nicotinamide ring of the coenzyme.

5. If you wanted to follow the course of the reaction in the previous question by spectrophotometric techniques, how could it be done? Why would this be more difficult than following the reverse reaction in the same manner?

6. Your text indicates that K_m is the substrate concentration that gives a rate of $1/2\ V_{max}$. Show that a substrate concentration of $2\ K_m$ does not give a rate of V_{max}. Do substrate concentrations of either $3\ K_m$ or $4\ K_m$ give a rate equal to V_{max}?

7. Why is the V_{max} of an enzyme-catalyzed reaction the same in the presence of a competitive inhibitor as it is in the absence of one?

8. How can one determine initial rates of reaction required for making Michaelis-Menten plots?

9. The K_m for the substrate of a particular enzyme is 2.0×10^{-5} M. If the initial velocity, v_o, is 0.16 µmol/min for [S] = 0.15, what will be the initial velocity when [S] = 5.0×10^{-3}?

10. Is a synthase the same as a synthetase?

11. Speculate on the advantage(s) for the cell of maintaining intracellular substrate concentrations within the range of $0.1\ K_m$ to K_m.

12. The following kinetic data were collected for an enzyme-catalyzed reaction. Determine (a) the V_{max} and (b) the K_m for this system.

Measurement Number	[S], M	v_o, µmol/min
1	1.0×10^{-5}	15.6
2	5.0×10^{-5}	34.6
3	1.0×10^{-4}	41.0
4	5.0×10^{-4}	47.9
5	6.0×10^{-2}	49.8
6	5.0×10^{-1}	50.0
7	8.0×10^{-1}	50.0

When a competitive inhibitor ($[I] = 1.4 \times 10^{-4}$) was added to this system with [S] = 1.0×10^{-4}, a 30.0% decrease in initial velocity was observed. Find (c) the K_i for this competitive inhibitor. Note: It has been determined that $K_m^{app} = K_m(1 + [I]/K_i)$.

13. Explain why trypsin, a protease, and α-amylase belong to the same IUBMB category of enzymes. Which category is it?

D. Additional Problems

1. Although it is in your text, see if you can derive the Lineweaver-Burk equation by simple manipulation of the Michaelis-Menten equation. Explain possible uses of the Lineweaver-Burk equation.

2. An Eadie-Hofstee plot is another graphical means once used to evaluate K_m and V_{max} for an enzyme-catalyzed reaction system. The equation used is:

$$v_o / [S] = V_{max} / K_m - v_o \, 1/ K_m$$

For fun, derive this equation from the Michaelis-Menten equation and describe the graph that would result from a plot of the appropriate enzymatic data.

3. In an enzyme-catalyzed reaction of the type:

$$2A \rightarrow B + C$$

substrate concentrations of 5.0×10^{-1}, 5.0×10^{-2}, or 25.0×10^{-2} mM all produced 30. millimoles of product C in a 12-minute period. Express the level of enzyme assayed by this system in terms of International Units. Note: An International Unit of enzyme activity is defined as the number of micromoles of substrate transformed per minute by the enzyme.

4. If 2.0 mL was the volume of the enzyme solution used in the previous problem and it contained 2.5×10^{-4} mg of total protein, what is the specific activity of the solution? Note: The specific activity of an enzyme preparation can be defined as the International Units of activity per mg of protein.

5. The enzyme in the previous two problems has a molecular weight of 12,500 g/mole and is thought to have just one active site per molecule. Based on the information in the previous two problems, what is the catalytic constant (turnover number) for this enzyme?

6. The kinetic data from an enzyme-catalyzed reaction is shown below. From this information, determine K_m and V_{max}.

Measurement Number	[S], M	v_o, μmol/min
1	1.7×10^{-6}	10.0
2	3.8×10^{-6}	20.0
3	1.2×10^{-5}	45.0
4	2.3×10^{-5}	60.0
5	8.5×10^{-5}	85.0

7. The enzymatic reaction of problem D.6. is studied again in the presence of an inhibitor at 3.5×10^{-5} M concentration, giving the data shown below. Determine the type of inhibition. Is either K_m or V_{max} changed by this inhibition? If so, explain how.

Measurement Number	[S], M	v_o, μmol/min
1	3.8×10^{-6}	12.2
2	1.2×10^{-5}	26.9
3	2.3×10^{-5}	36.8
4	8.5×10^{-5}	51.3

Mechanisms of Enzymes

I. KEY TERMS

With the help of this study guide, your textbook, and class notes, you should be able to define and explain the significance of the following terms:

acid-base catalysis	effective molarity	proximity effect
activation energy	electrophilic	site-directed mutagenesis
carbanion	free radical	transition state
carbocation	induced fit	transition-state analog
catalytic antibody	mechanism	transition-state stabilization
catalytic triad	nucleophilic	zymogen
covalent catalysis	nucleophilic substitution	
diffusion-controlled reaction	reaction	

II. EXERCISES

A. True-False

_____ 1.   A reactant species that is electron rich and attacks another reactant deficient in electrons is termed an electrophile.

_____ 2.   When a carbon-carbon bond is split such that one carbon atom loses both electrons, that carbon atom becomes a carbocation.

_____ 3.   A transition state proposed to form as part of a reaction mechanism is a transient species that cannot be trapped and studied.

_____ 4.   The interior of a globular protein is generally a hydrophobic region. The active site of an enzyme is usually a cleft or pit in the globular protein. All the amino acid residues in the active site, therefore, have hydrophobic side chains.

_____ 5.   Ping-pong kinetics is a hallmark of enzymes that catalyze reactions involving covalent catalysis.

_____ 6.   Relatively strong binding of reactants in the enzyme active site is necessary for efficient catalysis.

_____ 7.   Entropy of reactants increases as reactants are bound by enzymes.

_____ 8.   An enzyme active site binds more tightly to a transition-state analog than to its substrate.

_____ 9.   Chymotrypsin is a proteolytic enzyme with three α-amino termini (N-termini).

_____ 10.   Serine proteases contain the same amino acid residues in the catalytic triad that are part of their active sites.

_____ **11.** Enzymes typically have lower Km values for coenzymes than for small substrates.

_____ **12.** A free radical is a transient species that results from C-C bond cleavage that leaves a C with an unpaired electron.

B. Short Answers

1. A/an _____ is the species formed when bond cleavage leaves a carbon with both electrons.

2. The _____ is a description of the bond-breaking and bond-forming events in a chemical reaction.

3. The _____ is the term for an energized arrangement of atoms in which bonds are being formed and broken prior to product formation in an enzyme-catalyzed reaction.

4. The energy required for reactants to reach the transition state from their ground state is called the _____ of the reaction.

5. In a _____ reaction, every collision between reactant molecules gives rise to product.

6. Rate enhancement by the binding of reactants close to each other in the enzyme active site is called the _____.

7. Excessive stability of the enzyme-substrate complex, which is with substrate(s) tightly bound to the enzyme, is termed a/an _____.

8. Very strong hydrogen bonds that form when the electronegative atoms are less than 0.25 nm apart are termed _____.

9. In the catalytic triad of serine proteases, the side chain of a _____ residue acts as a covalent catalyst.

10. _____ is the term applied to the catalytic mode that involves increased binding of transition states to enzymes as compared with binding of substrates or products.

11. The four major modes of enzymic catalysis are: _____, _____, _____, and _____.

12. Activation of an enzyme by a substrate-initiated conformation change is called _____.

13. The most frequently found catalytic residue in the active site of enzymes is _____.

14. _____ are inactive enzyme precursors activated by removal of one or more portions of the precursor molecule.

C. Problems

1. In the reaction of a certain anionic species, $Y:^-$, with a reactant of the $R—C=O$ type, would you expect
   (with X below the C)

   (a) $Y:^-$ to attack the carbon or the oxygen atom? Explain.
   (b) the leaving group to be a cationic or an anionic species? Explain.

2. What is the difference between intermediates and transition states?

3. It is said that dehydrogenation is the most common form of biological oxidation. What species is removed from the molecule being oxidized? Why is this oxidation?

4. Show the primary structure, using appropriate abbreviations for the amino acids, of the products formed by action of the following enzymes and chemical agent on this nonapeptide.

Ala–Cys–Lys–Met–Phe–Arg–Ala–Tyr–Gly

(a) elastase   (b) trypsin
(c) chymotrypsin   (d) BrCN

5. Describe in a sentence or two how an enzyme causes a specific reaction to proceed more rapidly than the same uncatalyzed reaction.

6. If the serine proteases discussed in this chapter (trypsin, chymotrypsin, and elastase) contain the same catalytic triad, why do they catalyze cleavage of only those peptide bonds in which the carbonyl group is donated by specific amino acid residues?

7. What is the function of enteropeptidase?

8. Suggest why trypsin can activate trypsinogen molecules to create more active trypsin.

9. Pancreatic trypsin inhibitor binds to and inactivates trypsin formed in the pancreas. Since trypsin inhibitor is itself a protein, why is it not digested by trypsin?

10. A certain proteolytic enzyme was inactivated with diisopropylfluorophosphate (DIFP). The inactive enzyme was subjected to enzymatic hydrolysis, which produced various peptides and the compound below. What does this evidence indicate about the enzyme?

$$(CH_3)_2CH-O-\overset{\overset{\displaystyle O}{\|}}{\underset{\underset{\displaystyle O-CH(CH_3)_2}{|}}{P}}-O-CH_2-\overset{\overset{\displaystyle NH_3^+}{|}}{CH}-COO^-$$

11. Summarize the way the catalytic triad of a serine protease achieves hydrolysis of a peptide bond. Consult Figure 6.27 in your text for help with this question.

12. Triose phosphate isomerase has active site residues glutamate and histidine that act as acid-base catalysts. What is unusual about the histidine side chain in this proposed mechanism?

13. When the active site Glu in triose phosphate isomerase was replaced with Asp by site-directed mutagenesis, was the catalytic rate altered? Explain.

D. Additional Problems

1. The maximum rate for an enzyme-catalyzed reaction, with substrate in excess, at 20.0°C is 2.52 μmol/min. At 30.0°C, the rate increases to 36.6 μmol/min. What is the activation energy, Ea, for this reaction? Hint: Use the Arrhenius equation, $K = e^{-Ea/RT}$, in its linear form,

$$\ln k = - E_a/R \times (1/T) + \ln A$$

where k is the rate constant for the reaction, A is the *frequency factor*, R is the gas constant (8.314

J/molK), and T is the Kelvin temperature.

2. Figure 6.1 is the pH profile for the proteolytic enzyme pepsin, which operates in the stomach in a rather acidic environment. Pepsin, like the HIV-1 protease, is an aspartic protease, with two aspartate side chains in the active site. What information does Figure 6.1 provide in terms of the mechanism of the action of pepsin? What is a somewhat unusual aspect of this mechanism?

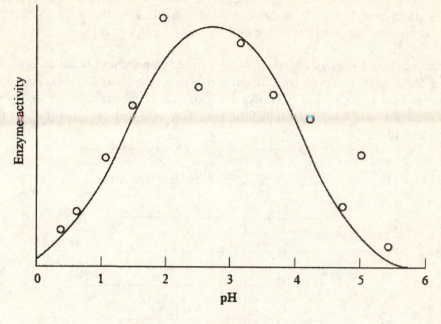

Figure 6.1 pH Profile of Pepsin

3. Figure 6.2 shows an example of burst kinetics for an enzyme catalyzed reaction of the type:

$$S \xrightarrow{\text{Enzyme}} P + Q$$

What mechanistic information is provided by this data?

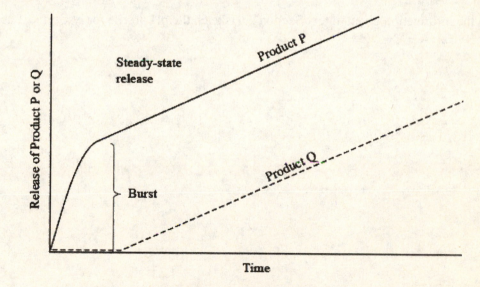

Figure 6.2 Burst Kinetics

4. For the nonenzymatic reaction shown here, there are 2.50×10^{-2} mmol of product R formed in a two minute period using 0.0100 M concentrations of A and B.

$$A + B \rightarrow R + T$$

When an enzyme is used to catalyze the reaction, using the same concentrations of A and B (and all other conditions also kept the same), there are 4.00 mmol of product R formed in a two-minute period. Evaluate the proximity effect provided by the enzyme by computing the effective molarity (Section 6.5A in your text), assuming all rate enhancement is due only to the proximity effect.

5. Lysozyme catalyzes the hydrolysis of the polysaccharide component of bacterial cell walls, which consists of alternating N-acetylglucosamine (NAG) and N-acetylmuramic acid (NAM) residues. Lysozyme also accepts chitin as a substrate. Chitin is a polysaccharide in crustacean shells that contains only NAG residues connected by β-1,4 linkages. The following data were collected in a study of lysozyme action on various oligomers of NAG.

Substrate	Relative Hydrolysis rate
NAG_2	0
NAG_3	1
NAG_4	8
NAG_5	4000
NAG_6	30,000
NAG_8	30,000

What conclusions can be drawn about the substrate specificity of lysozyme? Which particular glycosidic bond of the substrate is cleaved by lysozyme?

6. Based on the information given in text Section 6.6, sketch the pH profile for lysozyme. Label appropriate features.

Coenzymes and Vitamins

I. KEY TERMS

With the help of this study guide, your textbook, and class notes, you should be able to define and explain the significance of the following terms:

apoenzyme   group transfer protein   prosthetic group
coenzyme   holoenzyme   protein coenzyme
cofactor   iron-sulfur cluster   reactive center
cosubstrate   metal-activated enzyme
essential ion   metalloenzyme

II. EXERCISES

A. True-False

_____ 1. Iron-sulfur clusters are modified forms of heme groups involved in certain electron transfer processes.

_____ 2. Some coenzymes are made by an organism for its own use and are called metabolite coenzymes.

_____ 3. One should not take vitamin supplements that contain water-soluble vitamins, since an excess might lead to hypervitaminosis.

_____ 4. Pellagra is a disease characterized by dermatitis, impaired digestion, and diarrhea. Pellagra is associated with a deficiency of vitamin C.

_____ 5. Vitamin B_1, thiamine, deficiency is associated with the disease beriberi, which is characterized by rapid weight loss, muscle atrophy, and weakness.

_____ 6. Tetrahydrofolate, a coenzyme with a polyglutamate tail, is derived from the water-soluble vitamin folic acid.

_____ 7. Vitamin K is a lipid vitamin with antioxidant activity.

_____ 8. These reactants: $NADH + H^+ + Coenzyme Q$ are more likely to produce products than these reactants: $NAD^+ + Coenzyme QH_2$.

_____ 9. Cytochromes are heme-containing protein coenzymes in which the iron (III) ions of the heme undergo reversible one-electron reduction.

_____ 10. The four fat-soluble vitamins (A, D, E, and K) all function as coenzymes.

_____ 11. Both NAD^+ and $NADH$ absorb at 340 nm.

_____ **12.** Coenzyme A contains a structural component called ribitol.

_____ **13.** In an electric field, NAD^+ would migrate toward the anode.

B. Short Answers

1. Cofactors can be classified in one of two major categories, which are _____ and
_____.

2. Some enzymes do not require cofactors to be active. For those that do, the inactive protein itself, without
its cofactor, is called the _____.

3. When the inactive protein is combined with its specific cofactor, an active form is created called the
_____.

4. The function of coenzymes is to act as _____.

5. The first coenzymes to be recognized were _____ and _____.

6. Mobile metabolic groups transferred by coenzymes are temporarily attached to the _____ of
the coenzyme.

7. A coenzyme that is chemically altered during an enzyme-catalyzed reaction and that dissociates from the
active site is called a/an _____.

8. A coenzyme that does not dissociate from the active site but remains firmly bound to the enzyme is called
a/an _____.

9. A molecule with opposite ionic charges on adjacent atoms is called a/an _____.

10. The two main categories of vitamins are the _____ vitamins and the _____
vitamins.

11. The _____ are proteins that act as coenzymes. They usually contain such reaction centers as
_____, _____, and _____.

12. The mobile metabolic group of UDP-glucose is the _____.

13. _____ and _____ almost always act as cosubstrates for dehydrogenases.

14. _____ is the reducing agent used in biosynthetic (anabolic) reactions.

C. Problems

1. Explain the difference between metal-activated enzymes and metalloenzymes.

2. Explain why coenzymes are required by some enzymes.

3. For each of the following coenzymes, indicate a typical function or role:

   Coenzyme A _____

   NAD^+ _____

   Biotin _____

Thiamine pyrophosphate _____

Tetrahydrofolate _____

ATP _____

FAD _____

4. Which vitamin-derived coenzyme exists as a lactone? What is a lactone? What is the function of this coenzyme?

5. What group is the reactive center of acetyl CoA?

6. The reduced form of FAD is $FADH_2$. Some older texts also showed the reduced form of NAD^+ as $NADH_2$. Discuss the currently accepted mechanism for the reduction of NAD^+ and how it does not justify the use of the representation $NADH_2$ as the reduced form of this coenzyme.

7. Describe the sequence of steps in the addition of substrate and cosubstrate and the formation of product(s) in the oxidation of pyruvate to lactate by lactate dehydrogenase. Is this a ping-pong mechanism?

8. A histidine and an arginine residue are known to play roles in the active site of lactate dehydrogenase. Explain their involvement.

9. Both NAD^+ and FAD (also FMN) are coenzymes for dehydrogenase enzymes. They differ not only in the manner in which they are held by the enzymes, but also in the way that electrons are transferred. Explain the differences and show the mechanisms by which NAD^+ and FAD are respectively reduced.

10. What mammalian coenzyme has a long hydrophobic chain of isoprenoid units and what is the function of this chain?

11. What is the reduction potential of a cytochrome and why does it vary from one cytochrome to another even if the cytochromes of a given class (*a*-cytochromes, *b*-cytochromes, *c*-cytochromes) have the same heme groups?

12. Carbonic anhydrase is activated by Zn^{2+}. What is the function of Zn^{2+}?

13. Cite a statement in your text that is an indicator of the importance of S-adenosylmethionine.

14. You have a bottle of a substance that is either FAD or $FADH_2$. The substance is bright yellow. Which form of the coenzyme is in the bottle?

15. Cobalamine is synthesized by only a few microorganisms, so how do humans satisfy their requirement of this micronutrient?

D. Additional Problems

1. Pyruvate carboxylase is an enzyme that catalyzes the carboxylation of pyruvate to form oxaloacetate. The enzyme requires biotin as a coenzyme. ATP is also needed. Write the mechanism for this enzymatic reaction.

2. How does inhibition of the enzyme dihydrofolate reductase help retard the growth of cancer cells?

3. In one of the "Rocky" movies, Sylvester Stallone drank a concoction that contained raw eggs. He saved time with respect to cooking, but what health problem might arise from consuming raw eggs too often? What coenzyme is involved?

4. (a) What coenzyme forms Schiff bases? (b) What are aldimines? (c) What is the difference between

internal and external aldimines with respect to the mechanism of transamination?

5. Although somewhat rare, pernicious anemia may develop in some older people and in long-term strict vegetarians. (a) What coenzyme is involved/affected in pernicious anemia, and (b) what are some unusual aspects of this coenzyme? (c) What is possibly different in the development of this disease in older people as opposed to its development in the long-term vegetarians?

Carbohydrates

I. KEY TERMS

With the help of this study guide, your textbook, and class notes, you should be able to define and explain the significance of the following terms:

aglycone
aldonic acid
aldose
alduronic acid
amylopectin
amylose
anomer
anomeric carbon
epimer
furanose
glucoconjugate

glucoside
glycan
glycoprotein
glycosaminoglycan
glycoside
glycosidic bond
heteroglycan
homoglycan
ketose
limit dextrin
monosaccharide

N-linked oligosaccharide
oligosaccharide
O-linked oligosaccharide
peptidoglycan
polysaccharide
proteoglycan
pyranose
reducing sugar
sugar alcohol
triose

II. EXERCISES

A. True-False

_____ 1.   Dihydroxyacetone could be described as a ketotetrose.

_____ 2.   The naturally occurring sugars, like the natural amino acids, have a stereochemistry that places them in the L-series of sugars.

_____ 3.   Epimers are two carbohydrates that differ only in the configuration of groups at a single chiral center.

_____ 4.   Monosaccharides that form six-membered rings are called pyranoses.

_____ 5.   The complete hydrolysis of starch, glycogen, and cellulose would yield only D-glucose.

_____ 6.   Amylopectin is an entirely linear polymer of D-glucose units found in natural starch.

_____ 7.   Lactose is a disaccharide that contains a β-(1→4) type linkage.

_____ 8.   Maltose, lactose, sucrose, and cellobiose are all reducing sugars.

_____ 9.   Glycogen is like the amylopectin of starch except that glycogen has β-(1→4) linkages between D-glucose units.

_____ 10.   Fructose forms a cyclic structure that is a hemiketal.

_____ 11.   D-Glucose and D-galactose are epimers.

_____ 12. D-Mannuronic acid is the product of the oxidation of the C-1 aldehyde group of D-mannose.

_____ 13. Glycerol is a sugar alcohol.

_____ 14. The Gram stain procedure for bacteria involves interaction of a purple dye with peptidoglycans of the bacterial cell walls.

B. Short Answers

1. A carbohydrate that is made by joining five monosaccharides is called a/an _____.

2. The new chiral center formed when a monosaccharide such as D-galactose assumes a cyclic form is called the _____.

3. The two different forms of D-fructose that exist in solution, which differ in the –OH group orientation at the number 2 carbon, are called _____.

4. The nonsugar portion of a glycoside is called a/an _____.

5. The digestion of starch by amylase produces an amylase-resistant core called a/an _____.

6. The oxidation of C-1 of D-galactose to –COOH yields a compound called _____.

7. The homoglycan _____, composed of _____ units in β-(1→4) type linkages, is found in the exoskeletons of insects and crustaceans.

8. Bacterial cell walls contain a heteroglycan that is classified as a/an _____.

9. Large glycoproteins found in mucus, which contain _O_-linked oligosaccharides, are called _____.

10. An epimer of D-fructose is _____.

11. Mannitol is an example of a type of compound called a/an _____.

12. Of the four common disaccharides lactose, sucrose, cellobiose, and maltose, _____ and _____ are epimers.

C. Problems

1. Draw the Fischer projection of the enantiomer of D-galactose.

2. Draw the structure of 2-deoxy-α-D-ribofuranose.

3. If approximately 4% of the glucose units in amylopectin are involved in α-(1—>6) linkages where branching occurs, how many branches would there be on a linear portion of amylopectin that had 212 glucose units in that linear portion?

4. Explain why sucrose is not a reducing sugar when it is composed of glucose and fructose, both of which are reducing sugars.

5. List the monomeric units that make up the following:

a. amylopectin _____

b. chitin _____

c. milk sugar_____

d. maltose _____

e. glycogen_____

6. Draw the structure of β-D-mannopyranosyl-(1→5)-α-D-ribofuranose. Is this a reducing sugar?

7. It is possible to reduce the aldehyde group of aldoses like glucose, mannose, and ribose to an alcohol group. These derivatives are called sugar alcohols. Some (xylitol) have been used in sugarless gums (Orbit™) and other similar products. Draw the structures of the sugar alcohols glucitol and ribitol. Will they exist in cyclic or open chain forms? Explain.

8. Assume that glycogen has a molar mass of about 3.0×10^6 g. (a) How many individual glucosyl residues does it contain? (b) How many of these are located at branch points?

9. If about a pound of glycogen can be stored in the body of an adult male, how many molecules of the size indicated in the previous problem are present in this amount of glycogen?

10. (a) The linear Fischer projection form of an aldohexose has how many chiral carbons? (b) Based on this, how many isomers are possible? Compare your answer with the number of aldohexoses shown in text Figure 8.3. (c) What do you conclude?

11. Make the same comparison based on the Haworth projection form of an aldohexose. What do you conclude?

12. Methyl β-D-galactopyranoside is what kind of compound? Is it a reducing compound? Does it show mutarotation?

13. Of the disaccharides maltose, lactose, sucrose, and cellobiose, which one(s) can exist in anomeric form?

14. (a) Consider a molecule of amylose of mass 3.0×10^6 g. How many reducing ends does it have? (b) Answer the same question for a molecule of amylopectin of the same mass.

D. Additional Problems

1. Because it contains a number of chiral centers, D-glucose is optically active. It has been found, however, that a solution of the pure α-anomer has an initial "specific rotation" value different from that of a solution of the pure β-anomer. The initial value for α-D-glucose is +112.2°, while that for β-D-glucose is +18.7°. In both cases, these values change with time (that for the α-anomer decreases and that for the β-anomer increases) until an equilibrium value is reached. Use the percentage of each anomer present in aqueous solution as indicated in your text and calculate the equilibrium value of the specific rotation for glucose. Disregard possible contribution of any open chain form.

2. The specific rotation, [α], of a compound is related to the observed optical rotation of a solution of that compound as indicated:

$$[\alpha] = A/lc$$

where A is the observed optical activity in degrees, l is the cell path length in decimeters, and c is the concentration in grams per milliliter. Sucrose is dextrorotatory with a specific rotation of about +66.5°. A sucrose solution measured in a 25 cm polarimeter tube showed an optical rotation of +46.8°. What is the concentration of the sucrose solution?

3. An aqueous solution of β-D-galactopyranose initially exhibits a specific rotation of about +53° but, over time, reaches a value of about +80°. An aqueous solution of α-D-galactopyranose exhibits an initial specific rotation of +151°, but finally reaches an equilibrium value of +80°. Disregard any open chain form and calculate the percentage of α– and β– forms in the equilibrium mixture.

4. Some sources indicate that amylose forms a helix in which there are six glucosyl residues per turn of the helix and that each turn occupies a linear distance of 0.8 nm. (a) How long is a helical segment of amylose that contains 900 glucosyl residues? (b) How does this helix compare with the α-helix of proteins?

5. Reducing sugars, such as D-galactose, react with Ag$^+$ in aqueous ammonia causing elemental silver to be deposited on the walls of the container. The reaction is known as the Tollen's silver mirror test. The open chain form of the sugar is required for a positive test. (a) Write the equation for this reaction with D-galactose and name the sugar derivative formed. (b) If less than 1% of the open chain form of the sugar is normally present in aqueous solution, how can this reaction occur to any measurable extent?

6. Research suggests that glucose may be involved with the aging process as well as some of the complications associated with diabetes. Researchers think that the open chain form of glucose reacts with amine groups of proteins to first form a Schiff base that rearranges to an Amadori product (a keto derivative formed from a Schiff base by two tautomeric shifts of hydrogen atoms). The Amadori product (an N-substituted 1-amino-1-deoxy-2-ketose) is thought to react with another glucose and an amine group of another protein to form a cross link between the protein molecules. This type of cross link has been identified as 2-furanyl-4(5)-(2-furanyl)-1H-imidazole (FFI). It is suspected that cross links of this type cause collagen to become tough and inflexible. Other proteins such as those in the lens of the eye may also be affected, leading to cataract formation. (a) Use appropriate structures to illustrate the formation of a Schiff base with an amino group of a protein and its rearrangement to form an Amadori product. (b) Also illustrate the formation of the FFI cross link between two proteins.

For a little help: All carbons in the FFI cross link are supplied by the two glucose molecules, including the carbons of the two furanyl moieties and the imidazole moiety. One of the furanyl moieties is bridged to the imidazole ring by a carbonyl group, the carbon of which came from one of the glucose molecules. The two nitrogen atoms of the imidazole ring were supplied by the amino groups of the two protein chains that were cross linked.

7. How many different disaccharides can be made from D-glucopyranose and D-fructofuranose? (Remember that disaccharide linkages involve at least one anomeric carbon.)

Lipids and Membranes

I. KEY TERMS

With the help of this study guide, your textbook, and class notes, you should be able to define and explain the significance of the following terms:

active transport
antiport
cascade
ceramide
cerebroside
channel
chemotaxis
effector enzyme
eicosanoid
endocytosis
exocytosis
fatty acid
fluid mosaic model
freeze-fracture electron
ganglioside
glycerophospholipid
glycosphingolipid
G-protein

growth factor
hormone
integral membrane protein
isoprenoid
lateral diffusion
lipid
lipid bilayer
lipid raft
lipid-anchored   membrane
protein
microscopy
monounsaturated
neurotransmitter
passive transport
peripheral membrane protein
phospholipid
plasmalogen
polyunsaturated

pore
prenylated protein
primary active transport
prostaglandin
saturated
second messenger
secondary active transport
sphingolipid
sphingomyelin
steroid
sterol
symport
transducer
transverse diffusion
triacylglycerol
uniport
unsaturated
wax

II. EXERCISES

A. True-False

_____ 1. Double bonds in polyunsaturated fatty acids are conjugated.

_____ 2. Phosphatidyl choline (also called lecithin) is a type of phospholipid found in biological membranes.

_____ 3. Of the three types of membrane lipids mentioned in the text, all are charged.

_____ 4. Plasmalogens are sphingosine derivatives.

_____ 5. All phospholipases catalyze the specific hydrolysis of phosphoester bonds of phospholipids.

_____ 6. The fatty acid component of a ceramide is attached to sphingosine via an ester bond.

_____ 7. Gangliosides contain the N-acetylneuraminic acid group.

_____ 8. Micelles are aggregates of cholesterol.

_____ **9.** Membrane lipid bilayers are permeable to water.

_____ **10.** Triacylglycerols are neutral lipids.

_____ **11.** Membrane proteins are associated only with the surfaces of biological membranes.

_____ **12.** The two surfaces of a biological membrane differ from each other.

_____ **13.** Integral proteins can move within the lipid bilayer of the membrane.

_____ **14.** A wax is an ester made from a long chain alcohol and a long chain fatty acid.

B. Short Answers

1. _____ and _____ are essential fatty acids for mammals.

2. Three major types of lipids in animal membranes are _____, _____, and _____.

3. A moiety common in the head group of all phosphatidates is _____.

4. Three common aminoalcohols found in phosphatidates are _____, _____, and _____.

5. _____ are glycosphingolipids that provide cells with distinctive surface markers such as those that provide the basis for ABO blood-grouping.

6. _____ are glycosphingolipids abundant in nerve tissue.

7. The type of enzyme that catalyzes formation of phosphatidates from glycerophospholipids is _____.

8. _____ contain sphingosine, a fatty acid, and a carbohydrate such as glucose, galactose, or an oligosaccharide.

9. _____ is an ATP-dependent protein that moves phospholipids from the outer to the inner leaflet of a membrane.

10. The more fluid phase of membranes is called the_____, and the less fluid phase is called the _____.

11. The movement of a phospholipid from one monolayer to the other in a membrane is called_____.

12. The two types of proteins associated with membranes are the _____ and _____ proteins.

13. Specific peripheral proteins on the inner surfaces of membranes that have GTPase activity and function as signal transducers are called _____.

C. Problems

1. List (name) five of the diverse functions of lipids.

2. List the products of the complete hydrolysis of the following: (a) glucocerebroside, (b) sphingomyelin, and (c) lecithin.

3. How do cholesterol molecules orient themselves in biological membranes?

4. What are eicosanoids? Name three different types of eicosanoids that exhibit different biological functions.

5. What property of triacylglycerols makes them efficient energy storage molecules? Explain.

6. What is the driving force for the formation of lipid mono- and bilayers in/on an aqueous solution?

7. Name a type of membrane rich in protein.

8. How are carbohydrates arranged in membranes?

9. Why does treatment of biological membranes with chelating agents, such as ethylenediaminetetraacetic acid (EDTA), cause the release of peripheral proteins?

10. Explain the difference between primary and secondary active transport in membranes.

11. Explain how integral proteins are associated with the lipid bilayer.

12. Thin layer chromatography of a triacylglycerol and lecithin on silica gel in chloroform-methanol-water developing solvent shows that lecithin has an R_f value of about 0.4 and the triacylglycerol travels nearly with the solvent front. Why are they so different?

13. Apparently the fluidity of membranes is important to their biological functions. It was shown that in *E. coli* the ratio of saturated to unsaturated fatty acids decreases as the temperature decreases. How does the presence of unsaturated fatty acids affect the fluidity of membranes? What role does cholesterol play in mammalian membrane fluidity?

14. Briefly contrast the mode of lipid vesicle formation occurring in endocytosis with that occurring in exocytosis.

15. What is the difference between a membrane pore and a channel?

D. Additional Problems

1. If there are 2.86×10^4 phospholipid molecules in a section of lipid bilayer that has a surface area on each face of the membrane of 100.0 μm^2, what is the surface area occupied by a single molecule?

2. A liposome is a vesicle formed by dispersion of phospholipids in aqueous salt solutions. If such a liposome had a diameter of 40.0 nm, what is the volume of aqueous solution enclosed by the liposome? The thickness of a lipid bilayer is about 4–5 nm.

3. Based on the information in the previous problem, how many amino acid residues in an α-helical segment of a protein are required to span the membrane?

4. Studies have indicated that the diffusion constant, D, for a phospholipid in a membrane is about 10^{-8} cm^2/sec. The distance, s, (in cm) traveled by a molecule in t seconds is given by: $s = (4Dt)^{1/2}$. How long would it take for a lipid molecule to travel the length of a bacterium that is about 2.0 μm in length?

5. Membrane asymmetry is often studied by the use of 2,4,6-trinitrobenzene sulfonic acid (TNBS), which undergoes nucleophilic attack by the primary amine groups of certain lipids. What specific membrane lipids will react with TNBS? Write a generalized equation (just show the functional group on a "stick" lipid) for this reaction. Hint: Sulfite is liberated. For an additional challenge, write the mechanism for this reaction.

6. Bacteriorhodopsin is the only type of protein in the "purple membrane" of *Halobacterium halobium*. The membrane is about 4 nm thick and the protein contains 247 amino acid residues. Portions of the protein that traverse the membrane are arranged in the α-helix type of secondary structure. Only 27% of the amino acid residues of the protein are not involved in these α-helical areas. How many times does the bacteriorhodopsin chain traverse the membrane?

Introduction to Metabolism

I. KEY TERMS

With the help of this study guide, your textbook, and class notes, you should be able to define and explain the significance of the following terms:

anabolic reaction
autotroph
catabolic reaction
chemoautotroph
chemoheterotroph
citric acid cycle
electromotive force
energy-rich compound
feedback inhibition
feed-forward activation
flux
glycolysis

heterotroph
mass-action ratio, Q
metabolic fuel
metabolically irreversible
   reaction
metabolism
metabolite
near-equilibrium reaction
oxidative phosphorylation
oxidizing agent
pathway
phosphagen

phosphoryl-group-transfer
   potential
photoautotroph
photoheterotroph
protein kinase
protein phosphatase
reducing agent
reduction potential
standard state
turnover

II. EXERCISES

A. True-False

_____ 1. Metabolism is an inclusive term used to indicate all reactions carried out by living cells.

_____ 2. The enzymes of the glycolytic pathway are found in the cell cytoplasm. Since they are not compartmentalized, glycolysis is always "on" and will convert glucose to pyruvate as long as any glucose is available.

_____ 3. Cells are very efficient at capturing the stored energy in a molecule such as glucose; thus, the efficiency of energy conversion approaches 100%.

_____ 4. One gram of NaCl dissolved in a test tube of water has a greater entropy than does a gram of NaCl in the form of a crystalline solid.

_____ 5. The Gibbs free energy change is a measure of the energy available in a reaction.

_____ 6. For a reaction to be spontaneous, the change in free energy, ΔG, must be negative.

_____ 7. Processes that result in an increase of the entropy of the system are always spontaneous.

_____ 8. The free energy change, ΔG, in a system is dictated only by the absolute temperature and the change in heat content (enthalpy), ΔH, of the system.

_____ 9. Cellular reactions tend to reach equilibrium so that ΔG becomes zero.

_____ **10.** Cellular reactions for which the steady state ratio of products to reactants, Q, is numerically not close to the normal equilibrium constant, K_{eq}, are known as metabolically irreversible reactions.

_____ **11.** For a reaction of the type A + B $\leftrightarrow$ C + D, the value of the equilibrium constant ($K_{eq} = 0.125$) indicates that the reaction proceeds nearly to completion.

_____ **12.** In biological systems, ATP plays a central role in energy transfer since it has an intermediate phosphoryl transfer potential.

_____ **13.** Glycolysis is an example of a linear metabolic pathway.

B. Short Answers

1. _____ reactions are those by which the cell synthesizes needed materials such as proteins and nucleic acids, while _____ reactions are those by which the cell degrades materials for energy and building blocks.

2. Three "forms" of metabolic pathways are _____, _____ , and _____.

3. The flow of metabolites through a pathway is called _____.

4. Two events that increase the likelihood of a spontaneous process are a large _____ in the entropy of the system and a/the _____ of heat by the system.

5. The multistep catabolism of glucose to pyruvate is called _____.

6. The type of bond in ATP or ADP that has the highest free energy of hydrolysis is called a/an _____ .

7. A molecule or species that loses one or more electrons in a reaction is called a/an _____ agent and becomes _____.

8. Energy from catabolic pathways is conserved in the form of _____, _____, or _____.

9. Energy-rich phosphate-storage molecules that occur in muscle tissue are called _____. Two examples are _____ and _____.

10. Phosphorylated compounds with phosphoryl-group-transfer potentials equal to or higher than that of ATP are termed _____.

11. Metabolic pathways are controlled by the modulation of allosteric enzymes. The rate of the pathway can be slowed by the action of a/an _____ of the pathway, a process called _____. The pathway rate can be increased by the action of a/an _____, a process called _____.

12. Individual enzymes in a pathway are regulated by _____ or by reversible _____.

13. The relative tendency, in an oxidation-reduction reaction, for one species to accept electrons from another species (be reduced) is termed the _____.

14. A nonphotosynthetic organism that requires an organic carbon source such as glucose is called a/an _____ .

15. Two biological processes that are entropy driven are _____ and _____.

C. Problems

1. Your text indicates in one of the common themes of organisms that specific internal concentrations of enzymes and metabolites are maintained. In another of these common themes, your text indicates that there is a continual degradation and synthesis of cell components. Are these two statements in conflict? Explain.

2. Of the examples of metabolic pathway types given in text Figure 10.2, which are anabolic? Which are catabolic?

3. Can a reaction with a positive standard free energy change, $\Delta G^{\circ\prime}$, proceed in the forward direction? Explain.

4. For a reaction with a positive $\Delta G^{\circ\prime}$ value, what factor(s) would yield a negative ΔG such that the reaction would proceed spontaneously in the forward direction?

5. Using the information in text Table 10.3, determine the equilibrium constant, K_{eq}, for the following reaction at 25°C:

$$\text{glucose 1-phosphate} + H_2O \rightarrow \text{glucose} + P_i$$

6. In addition to regulation of pathways by allosteric control, pathway flux may also be controlled by covalent modification of interconvertible enzymes, such as by the addition or removal of phosphoryl groups. Are catabolic enzymes activated or deactivated by phosphorylation? Are anabolic enzymes activated or deactivated by phosphorylation?

7. Cite two or three consequences of the fact that various metabolic processes are sequestered in specialized subcellular compartments.

8. The *in vitro* oxidation of glucose to CO_2 and H_2O proceeds with a $\Delta G^{\circ\prime}$ of -686 kcal/mol. The *in vivo* oxidation of glucose results in the formation of 32 moles of ATP from ADP and P_i. (See text Figure 13.10.) Determine the percentage of the "available" energy in glucose that is captured as ATP by cellular metabolism. The value for the total moles of ATP, 32, is obtained by tabulating all substrate level phosphorylation in glycolysis and the citric acid cycle as well as that formed by oxidation of all reduced cofactors formed (NADH and QH_2) via the electron transport system.

9. The hydrolysis of glucose 6-phosphate to glucose and P_i has a $\Delta G^{\circ\prime}$ value of -14 kJ/mol. If the steady state concentrations in a certain tissue of glucose 6-phosphate, glucose, and P_i were 10^{-3} M, 2×10^{-4} M, and 5×10^{-2} M, respectively, what is the change in free energy, ΔG, for the hydrolysis at normal body temperature of 37°C?

10. Imagine, in a certain biochemical pathway, that the standard free energy, $\Delta G^{\circ\prime}$, for the formation of a thioester, R–CO–SR', from the reagents R–COOH and R'–SH is 10. kJ/mol. (a) Use the information below to propose a two-step process by which the cell might spontaneously form the thioester. (b) Calculate the actual free energy change for the two-step process. Should this process proceed spontaneously? Explain.

$$ATP + H_2O \rightarrow AMP + PP_i \quad \Delta G^{\circ\prime} = -32 \text{ kJ/mol}$$

11. Your text illustrates the calculation of the standard free energy change, $\Delta G^{\circ\prime}$, for the oxidation of NADH by the electron transport chain with oxygen as the ultimate acceptor of electrons. Do the same kind of calculation for the oxidation of succinate by Q (ubiquinone).

12. The first reaction of the citric acid cycle, the condensation of oxaloacetate with acetyl CoA, has a standard free energy change of -32.2 kJ/mol. The last reaction in the cycle, regeneration of oxaloacetate from malate, has a standard free energy change of 29.2 kJ/mol. Explain how the cycle can continue to operate when the reaction that regenerates oxaloacetate has a positive free energy change.

13. Consider the reduction of pyruvate to lactate by NADH. Calculate the standard free energy change, $\Delta G^{\circ\prime}$, for this process. (You may need to use text Table 10.4.)

D. Additional Problems

1. Fully protonated ATP could be written as H_4ATP. However, ATP is extensively ionized at physiological pH. Consider this equilibrium:

$$HATP^{3-} \leftrightarrow H^+ + ATP^{4-} \quad pK_a = 6.95$$

(a) Barring any other influences, what would be the percentage of the fully ionized form of ATP in blood of normal pH 7.4? (b) How does this differ from the percentage of fully ionized ATP in cells of pH 7.0?

2. In the glycolytic pathway, phosphorylation of free glucose is catalyzed by hexokinase with ATP as the phosphoryl-group donor. (a) If the $\Delta G^{\circ\prime}$ is -16.7 kJ/mol for this reaction, what is the value of the equilibrium constant at 37°C? (b) If the initial concentrations of glucose and ATP were each 0.0500 M, what would be the equilibrium concentrations of all species? (Note: This hypothetical problem may not represent actual cellular concentrations of glucose and ATP and it also supposes that an equilibrium could be reached. Actually, the product glucose 6-phosphate is immediately used in the next reaction of the pathway to maintain a steady state of flux through the pathway.)

3. In erythrocytes, the steady state concentration of ATP is estimated to be 13 times that of ADP. The standard free energy change, $\Delta G^{\circ\prime}$, for glucose phosphorylation catalyzed by hexokinase is -16.7 kJ/mol, while the actual free energy change, ΔG, is -33.9 kJ/mol. Under these conditions, what would be the ratio of glucose to glucose 6-phosphate at 37°C?

4. In a cell at 37°C, the actual free energy change, ΔG, for the hydrolysis of ATP to ADP and P_i was determined to be -41.8 kJ/mol. Consider the standard free energy change, $\Delta G^{\circ\prime}$, given for this reaction in text Table 10.1 and estimate the ratio of ATP to ADP in this cell. Also indicate whether the hydrolysis of ATP is spontaneous under these conditions.

5. Some microorganisms form lactate from pyruvate (the product of glycolysis); this also occurs in exercising muscle. One source gives the standard free energy change for the net conversion of glucose to lactate as -123.5 kJ/mol. The overall reaction is:

$$glucose + 2\ ADP + 2\ P_i \rightarrow 2\ lactate + 2\ ATP$$

The normal glycolytic conversion of glucose to pyruvate has a standard free energy change of about -73 kJ/mol according to this net reaction:

$$glucose + 2\ ADP + 2\ P_i + 2\ NAD^+ \rightarrow 2\ pyruvate + 2\ ATP + 2\ NADH + 2\ H^+$$

Contrast these two systems in terms of (1) energy production, (2) extent of oxidation of glucose, and (3) relative likelihood of the reactions to proceed spontaneously as written.

Glycolysis

I. KEY TERMS

With the help of this study guide, your textbook, and class notes, you should be able to define and explain the significance of the following terms:

glycolysis   obligatory glycolytic tissues   substrate-level phosphorylation

isozyme   Pasteur effect

II. EXERCISES

A. True-False

_____ 1. Glycolytic enzymes are found in mitochondria.

_____ 2. Overall, glycolysis is an energy-releasing (exergonic) process.

_____ 3. Glycolysis involves the anaerobic oxidation of glucose.

_____ 4. Glycolysis produces no reduced cofactors.

_____ 5. In glycolysis, there are two intermediates that contain two phosphate groups each. The other intermediates have only one phosphate group.

_____ 6. Four of the ten reactions of glycolysis are catalyzed by kinases.

_____ 7. Alcoholic fermentation in yeast duplicates the glycolytic process except for two final steps.

_____ 8. Humans have all the enzymes but one necessary for alcoholic fermentation.

_____ 9. Reaction eight of glycolysis, the conversion of 2-phosphoglycerate to 3-phosphoglycerate, is catalyzed by phosphoglycerate mutase. Since the product of the reaction is an isomer of the substrate, the enzyme is classified as an isomerase.

_____ 10. Glycolysis consists of two stages, the hexose stage and the triose stage. A difference is that the hexose stage requires ATP and the triose stage produces ATP.

_____ 11. The third reaction of glycolysis is regulated in addition to the first reaction of glycolysis, since substrates can enter the pathway as fructose 6-phosphate, bypassing the first reaction.

_____ 12. The action of triose phosphate isomerase with dihydroxyacetone phosphate as substrate forms equal amounts of D- and L-glyceraldehyde-3-phosphate.

_____ 13. In mammals, free glucose enters cells under the control of insulin and with the aid of passive transporters called GLUT 1, 2, 3, and so on. Glucose remains in its free form until ATP levels are low and it enters the glycolytic pathway.

_____ **14.** Fructose, mannose, and galactose are all catabolized by glycolysis after conversion to appropriate intermediates of the pathway. The moles of ATP produced by glycolysis are the same for all of these hexoses.

B. Short Answers

1. Pyruvate kinase occurs as _____, different forms of the enzyme that catalyze the same reaction, but may have different Km values for the substrate, _____.

2. Enzymes that catalyze the hydrolytic removal of a phosphoryl group from a substrate are called _____.

3. Pyruvate kinase isozymes are allosterically activated by _____ and inhibited by _____.

4. Three glycolytic enzymes under metabolic control are _____, _____, and _____.

5. In the aldolase-catalyzed cleavage of fructose 1,6-*bis*phosphate, carbons 1, 2, and 3 of the substrate are found in the product _____.

6. The enzyme-catalyzed formation of ATP or some other high-energy nucleotide directly in a pathway step is called _____.

7. An enzyme that catalyzes the conversion of one sugar phosphate into another by changing the orientation of a single –OH group is called a/an _____.

8. The slowing of glycolysis in the presence of oxygen is called _____.

9. Lactose intolerance, a condition prevalent in adults in most parts of the world, is due to _____.

10. Two negative allosteric modulators of the enzyme phosphofructokinase-1 are _____ and _____.

11. The product(s) of the glycolytic step catalyzed by pyruvate kinase is(are) _____.

12. In erythrocytes, about _____ percent of the glycolytic flux is diverted for the formation of 2,3-*bis*phosphoglycerate.

13. The catabolism of which sugar—fructose, mannose, or galactose—requires the action of an epimerase? _____

C. Problems

1. Use Figure 11.1 to write the net reaction for the glycolytic formation of pyruvate from glucose. Structures are not necessary.

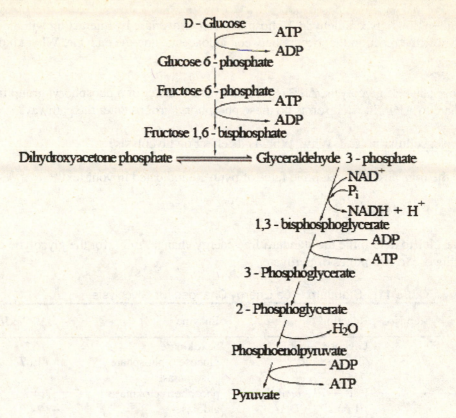

Figure 11.1 The Glycolytic Pathway

2. Assuming you had in solution all the <u>reactants</u> of the net reaction written in problem C.1. and all the glycolytic enzymes, would glycolysis proceed *in vitro* (in a test tube)?

3. Knowing that two molecules of ATP form for each glucose molecule metabolized via glycolysis, explain why your text indicates that the main glycolytic energy gain is due to the production of NADH in step 6.

4. Is lactate constantly formed in mammalian cells? Explain.

5. Fill in the blanks to show the initial glycolytic intermediates formed when each hexose is catabolized. Also indicate the number of moles of each intermediate formed per mole of each hexose.

Hexose	Initial Glycolytic Intermediate	Number of moles of Intermediate
mannose		
fructose		
galactose		

6. Of what importance is the function of the enzyme triose phosphate isomerase in glycolysis?

7. Write the equation for the overall net reaction that would occur if glyceraldehyde-3-phosphate were the input material for glycolysis.

8. Illustrate the cyclic process by which NAD^+ is regenerated to keep glycolysis going. Not all intermediates need to be shown. How is NAD^+ regeneration achieved in fermentation?

9. During glycolysis, what intermediate(s) accumulate(s) if an enolase inhibitor is present?

10. What would be the effect in an individual in whom the enzyme triose phosphate isomerase was either deficient or inhibited?

11. Iodoacetate inhibits glyceraldehyde 3-phosphate dehydrogenase by interacting with an active site cysteine residue. Show the reaction that occurs between iodoacetate and the enzyme. What kind of inhibition is this?

12. In the triose stage of glycolysis, ATP is formed by the transfer of a phosphoryl group from a glycolytic intermediate to ADP. In what forms do these phosphoryl groups enter the pathway?

13. Arsenate is a cellular poison. Why? What is its effect on glycolysis?

14. What are the four possible metabolic fates of pyruvate indicated in your text?

D. Additional Problems

1. Use Table 11.1 to determine the standard free energy change, $\Delta G^{\circ\prime}$, for the glycolytic conversion of one mole of glucose to two moles of pyruvate.

Table 11.1 Standard Free Energy Changes for Glycolysis

Step No.	Reaction	Enzyme	$\Delta G^{\circ\prime}$, kJ/mol
1	Glucose → G6P	hexokinase	-16.7
2	G6P → F6P	glucose 6-phosphate isomerase	+1.67
3	F6P → F1,6bisP	phosphofructokinase-1	-14.2
4	F1,6bisP → DHAP + G3P	aldolase	+24.0
5	DHAP → G3P	triose phosphate isomerase	+7.66
6	G3P → 1,3bisPG	glyceraldehyde 3-phosphate dehydrogenase	+6.28
7	1,3bisPG → 3PG	phosphoglycerate kinase	-18.8
8	3PG → 2PG	phosphoglycerate mutase	+4.44
9	2PG → PEP	enolase	+1.84
10	PEP → pyruvate	pyruvate kinase	-31.4

2. From the standpoint of the energy available from conversion of glucose to pyruvate by glycolysis, determine the percent efficiency of glycolysis based on the energy captured as ATP.

3. In exercising muscle, some of the glycolytic product pyruvate is converted to lactate. Is the efficiency of glycolysis increased or decreased by this additional step? Determine the percent efficiency for the production of two moles of lactate from one mole of glucose by glycolysis. The conversion of pyruvate to lactate is accompanied by a $\Delta G^{\circ\prime}$ of -25.1 kJ/mol.

4. The standard free energy change, $\Delta G^{\circ\prime}$, for the cleavage of fructose 1,6-bisphosphate into dihydroxyacetone phosphate and D-glyceraldehyde 3-phosphate (step 4) is +24 kJ/mol. With this large positive standard free energy change, how can glycolysis continue?

5. An enolase deficiency has an adverse effect on erythrocytes, specifically with respect to the delivery of oxygen to the tissues. Explain.

Gluconeogenesis, the Pentose Phosphate Pathway, and Glycogen Metabolism

I. KEY TERMS

With the help of this study guide, your textbook, and class notes, you should be able to define and explain the significance of the following terms:

Cori cycle	glucose-alanine cycle	limit dextrin
epinephrine	glycogenin	pentose phosphate pathway
glucagon	glycogenolysis	phosphorolysis
gluconeogenesis	insulin	substrate cycle

II. EXERCISES

A. True-False

_____ 1. In contrast to glycolysis, the pentose phosphate pathway allows the complete oxidation of glucose to CO_2.

_____ 2. The formation of DNA and RNA directly depends on high gluconeogenesis activity in the cell.

_____ 3. Glucose 1-phosphate is the direct product of glycogenolysis.

_____ 4. Glycogen synthase catalyzes the linkage of all glucose molecules used in the formation of glycogen in liver and muscle tissue.

_____ 5. Glycogen synthase catalyzes the joining of two glucose units, supplied as UDP-glucose, to initiate the formation of a glycogen chain.

_____ 6. Gluconeogenesis is simply a reversal of glycolysis that occurs when blood glucose levels fall below normal.

_____ 7. The formation of one mole of glucose by gluconeogenesis from pyruvate requires the same amount of energy as that produced by glycolytic degradation of one mole of glucose to pyruvate.

_____ 8. The Cori cycle is a combination of glycolysis and gluconeogenesis.

_____ 9. The enzymes that catalyze the reactions of the pentose phosphate pathway are all found in the cytosol.

_____ 10. NADPH is produced in the nonoxidative stage of the pentose phosphate pathway.

_____ 11. Normally, the brain relies almost entirely on glucose for its energy needs.

_____ 12. Most glucose 6-phosphate produced in the liver from glycogenolysis is converted to free glucose for delivery to cells of other tissues.

_____ 13. Glucagon is a small peptide hormone that stimulates glycogenolysis by specifically targeting liver cells.

B. Short Answers

1. The coenzyme required for reductive biosynthesis (e.g., of fatty acids), which is produced by the pentose phosphate pathway, is _____.

2. A pentose phosphate pathway enzyme that catalyzes the transfer of a three-carbon unit from a ketose-phosphate to an aldose-phosphate is called a/an _____.

3. Name the pathway or process discussed in text Chapter 12 to which each of the following belongs.

   pyruvate carboxylase _____

   sedoheptulose 7-phosphate _____

   glucose 6-phosphatase _____

   UDP-glucose _____

4. Glycogen phosphorylase catalyzes the degradation of glycogen chains from their nonreducing ends, but stops four glucose residues from a branch point. The remaining molecule is called a/an _____.

5. Gluconeogenesis requires four enzymes that are not enzymes of the glycolytic pathway. These four enzymes are _____, _____, _____, and _____.

6. The protein that is attached to the glycogen primer required for glycogen synthesis is called _____.

7. The principal hormones that control glycogen metabolism in mammals are _____, _____, and _____.

8. _____ is the process by which synthesis of PEP carboxykinase is increased due to increased transcription of the gene for this enzyme, triggered by increased levels of cAMP that result from prolonged release of glucagon.

9. Three major gluconeogenic precursors are _____, _____, and _____.

10. A pair of reactions that both form and degrade a specific substrate, in order to fine tune regulation of metabolism, is called a _____.

11. The primary organs/sites of gluconeogenesis in mammals are the _____ and _____.

C. Problems

1. What is the net energy produced/consumed by gluconeogenesis?

2. What is the net energy gain/loss in the conversion of two pyruvate molecules to one glucose molecule by gluconeogenesis if that glucose molecule is then catabolized to two pyruvate molecules by glycolysis?

3. In the first step of gluconeogenesis, is the C of CO_2 added to pyruvate to form oxaloacetate the same C of the CO_2 released in the next step as oxaloacetate is converted to phosphoenolpyruvate?

4. The liver is perfused in series with visceral tissues, while most other tissues are perfused in parallel. Of what consequence is this arrangement?

5. Figure 12.1 is the plot of kinetic data for the enzyme glycogen phosphorylase in the presence and absence of AMP. What can you conclude about this enzyme and the influence of AMP?

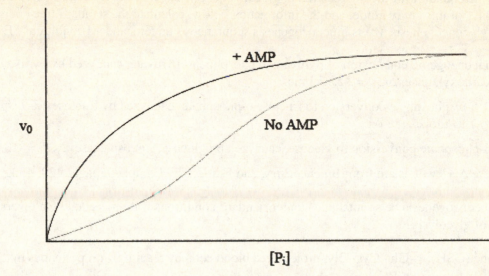

Figure 12.1 Effect of AMP on Glycogen Phosphorylase with Constant Glycogen Concentration

6. The gluconeogenic pathway uses four enzymes to bypass three reactions of glycolysis. Why are these bypass reactions necessary?

7. Contrast glycogenolysis in muscle and liver.

8. Glycogen is a highly branched polymer of glucose residues. Briefly explain how branches are formed in glycogen.

9. Your text says that "The level of PEP carboxykinase (PEPCK) activity in cells influences the rate of gluconeogenesis." (a) Using appropriate structures, write the equation for the reaction catalyzed by PEP carboxykinase. (b) How is the level of PEP carboxykinase controlled?

10. (a) What is the Cori cycle? (b) What is its function?

11. (a) In what tissues is the pentose phosphate pathway usually most active? (b) What is the cellular location of the pathway? (c) What are the principal functions of the pathway?

12. (a) What reaction is catalyzed by the enzyme 6-phosphogluconate dehydrogenase? (b) In what pathway does this reaction occur?

13. Write the names of the expected products in each of the following.

   (a) glyceraldehyde 3-phosphate + sedoheptulose 7-phosphate $\xrightarrow{\text{transketolase}}$

   (b) fructose 6-phosphate + glyceraldehyde 3-phosphate $\xrightarrow{\text{transketolase}}$

   (c) erythrose 4-phosphate + fructose 6-phosphate $\xrightarrow{\text{transaldolase}}$

14. Explain the connection between the pentose phosphate pathway and hemoglobin synthesis.

D. Additional Problems

1. (a) Using the equation for the net reaction of gluconeogenesis shown in text Section 12.1, the information in Table 11.1 of this study guide, and the information below, calculate the standard free energy change, $\Delta G^{\circ\prime}$, for gluconeogenesis. (b) Is gluconeogenesis spontaneous under standard conditions? Explain.

   <u>Given</u>: Gluconeogenic conversion of pyruvate to phosphoenolpyruvate, catalyzed by pyruvate carboxylase and PEP carboxykinase: $\Delta G^{\circ\prime} = +2.1$ kJ/mol.

   Fructose 1,6-*bis*phosphate conversion to fructose 6-phosphate, catalyzed by fructose 1,6-*bis*phosphatase: $\Delta G^{\circ\prime} = -16.7$ kJ/mol.

   Glucose 6-phosphate conversion to glucose, catalyzed by glucose 6-phosphatase: $\Delta G^{\circ\prime} = -13.8$ kJ/mol.

   Note: These $\Delta G^{\circ\prime}$ values are from *Biochemistry*, 2nd Edition, by Campbell, pages 429, 430.

2. Would gluconeogenesis be spontaneous under standard conditions if it proceeded by a reversal of all the reactions of glycolysis?

3. Glutathione (GSH), γ–Glu–Cys–Gly, protects red blood cells by reacting with peroxides that can cause
   |
   SH
   degradation of fatty acids in cell membranes and by reducing methemoglobin. In so doing, glutathione becomes oxidized to glutathione disulfide (GSSG). Glutathione is regenerated by the action of the enzyme glutathione reductase, which requires NADPH as a coenzyme. About 11% of African Americans have a deficiency of glucose 6-phosphate dehydrogenase that can lead to hemolytic anemia. (a) Using appropriate structures, write the reaction equation for the regeneration of glutathione from GSSG. (b) Glucose 6-phosphate dehydrogenase is an enzyme in what pathway discussed in this chapter? (c) Why does a deficiency of this enzyme possibly lead to hemolytic anemia?

4. Glycogen phosphorylase *a*, the active form, is allosterically inhibited by glucose and caffeine. How does this action of caffeine relate to the more familiar action of stimulation people usually experience from drinking coffee and other beverages that contain caffeine?

5. Suppose a particular glycogen molecule contained 6000 glucose residues. If a branch of eight residues occurred every eight residues of the glycogen polymer, (a) approximately how many chain ends would there be? (b) Are the ends of these chains reducing ends, nonreducing ends, or a mixture of the two?

6. How is glucose converted to fructose, the main fuel for sperm cells?

The Citric Acid Cycle

I. KEY TERMS

With the help of this study guide, your textbook, and class notes, you should be able to define and explain the significance of the following terms:

anaplerotic	glyoxylate cycle	pyruvate translocase
citric acid cycle	porin	

II. EXERCISES

A. True-False

_____ 1. FAD is used in the citric acid cycle as an oxidizing agent.

_____ 2. The final steps in the oxidation of glucose to CO_2 occur in the citric acid cycle.

_____ 3. Massive amounts of all intermediates of the citric acid cycle must be present for its efficient operation.

_____ 4. There are more six-carbon intermediates in the citric acid cycle than four-carbon intermediates.

_____ 5. In eukaryotes, the enzymes that catalyze the reactions of the citric acid cycle are found in the cytosol.

_____ 6. Pyruvate is produced in the eukaryotic cytosol but enters the mitochondrial matrix unaided.

_____ 7. The eukaryotic pyruvate dehydrogenase complex is the largest multienzyme complex known.

_____ 8. Intermediates of the citric acid cycle that are used for anabolic (synthetic) needs of the cell are replenished by anaplerotic reactions.

_____ 9. Porphyrin biosynthesis could potentially interfere with the citric acid cycle by depleting succinyl CoA.

_____ 10. Two moles of coenzyme A are consumed per mole of pyruvate catabolized to acetyl CoA, which is subsequently consumed in the citric acid cycle.

_____ 11. The citric acid cycle is controlled entirely by the availability of acetyl CoA and cycle intermediates.

_____ 12. Per mole of glucose, the citric acid cycle produces as many high-energy phosphate molecules by substrate level phosphorylation as does glycolytic catabolism of glucose to pyruvate.

_____ 13. A prochiral molecule is a symmetrical molecule that can be converted to a chiral molecule by substitution of a single functional group.

_____ **14.** The α-ketoglutarate dehydrogenase complex is very much like the pyruvate dehydrogenase complex.

_____ **15.** The reaction catalyzed by the succinate dehydrogenase complex is the only step in the citric acid cycle that involves substrate level phosphorylation.

B. Short Answers

1. The oxidation of pyruvate to acetyl CoA and CO_2 requires the cofactors _____, _____, _____, _____, and _____.

2. Four dicarboxylic acid intermediates of the citric acid cycle are _____, _____, _____, and _____.

3. The conversion of isocitrate to α-ketoglutarate also produces _____. This is an oxidative process in which the actual oxidant is _____.

4. Transformation of pyruvate to acetyl CoA is catalyzed by _____.

5. The enzyme that catalyzes the first step of the citric acid cycle is _____.

6. The enzyme that catalyzes the transformation of α-ketoglutarate to succinyl CoA is _____.

7. The only step in the citric acid cycle that directly produces an energy-rich phosphoanhydride involves the conversion of _____ to _____.

8. There are a total of _____ oxidation-reduction steps in the catabolism of pyruvate via the citric acid cycle. The specific reduced coenzymes formed are _____ and _____.

9. The competitive inhibitor _____ slows the citric acid cycle and causes succinate to accumulate.

10. Of the eight steps in the citric acid cycle, the number of steps that produce CO_2 is _____.

11. Certain citric acid cycle intermediates serve as biosynthetic precursors of other materials. Two such intermediates are _____ and _____.

12. A negative modulator involved in regulating each of the enzymes that catalyze the three metabolically irreversible reactions of the citric acid cycle is _____.

13. The intermediate _____ is formed by the hydration of _____ in the cycle.

14. The second CO_2 producing step in the citric acid cycle involves the conversion of _____ to _____.

C. Problems

1. What is coenzyme A and what is its function in the oxidation of pyruvate via the citric acid cycle?

2. List three sources of the acetyl group of acetyl CoA, a substrate of the citric acid cycle.

3. In eukaryotes, pyruvate is produced from glucose by glycolysis in the cytosol, but the enzymes of the citric acid cycle are in the mitochondrial matrix (except for succinate dehydrogenase, which is part of Complex

II in the inner mitochondrial membrane). How are these catabolic pathways connected despite being located in different cell compartments?

4. In the conversion of pyruvate to acetyl CoA, are all products released simultaneously? Explain.

5. In the aconitase reaction, how is a chiral product formed from citrate, which is not chiral?

6. How does the process of covalent modification control the pyruvate dehydrogenase complex?

7. Explain the effect of each of the following on the citric acid cycle in mammals and tell how that effect is achieved.
   (a) an abundance of ADP
   (b) a low supply of oxaloacetate

8. Determine the number of moles of each reduced cofactor and energy-rich phosphoanhydride (ATP and/or GTP) formed from the following as each is catabolized, in mammalian cells, through the stage of CO_2 generation via the citric acid cycle.
   (a) one mole of pyruvate
   (b) one mole of acetyl CoA
   (c) one mole of glucose

9. Write the overall (net) reaction for the conversion of isocitrate to succinate, which is accomplished by means of several steps in the citric acid cycle. Use appropriate structures for isocitrate and succinate.

10. If minced pigeon breast muscle is treated with malonic acid and oxaloacetate, what is the effect on the citric acid cycle? Would a specific cycle intermediate accumulate?

11. Aconitase catalyzes the conversion of citrate to isocitrate with *cis*-aconitate as an intermediate. At equilibrium *in vitro*, the relative concentrations are 90% citrate, 4% *cis*-aconitate, and 6% isocitrate. With so little isocitrate formed, how can the citric acid cycle function adequately *in vivo*?

12. Acetyl CoA modulates the activity of the pyruvate dehydrogenase complex. How does this allow the cell to respond to varying levels of acetyl CoA?

13. Using words and symbols (e.g., NAD^+), write the net reaction equation for the conversion of succinate to oxaloacetate by the citric acid cycle enzymes.

D. Additional Problems

1. A person was diagnosed as having a hereditary pyruvate dehydrogenase deficiency. How could the citric acid cycle operate adequately under such a circumstance? Should carbohydrate be restricted in this person's diet? Explain.

2. Some biochemistry texts state that glucose cannot be synthesized from fatty acids. A more definitive statement in one text is that "there can be no *net synthesis* of glucose from fatty acids, at least from the majority of fatty acids likely to be a part of the average diet." Oxaloacetate, in addition to being used for condensation with acetyl CoA to form citrate, can lead to glucose formation via gluconeogenesis. It has been demonstrated that, when labeled acetyl CoA from fatty acids is used in a functioning citric acid cycle, the label can be found in glucose. Explain how this can occur and yet be consistent with the "no net synthesis" statement above.

3. A derivative of pyruvate, hydroxypyruvate, must be converted to pyruvate for further catabolism. This seemingly simple change requires five separate reaction steps. The intermediates shown below are involved, as well as NAD^+, NADH, ATP, and ADP. Show the five reaction steps in a pathway using NAD^+, NADH, ATP, and ADP as needed. Hint: The last three steps of the sequence parallel those of glycolysis.

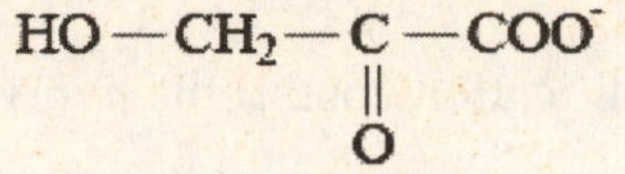

$$HO-CH_2-\underset{\underset{O}{\parallel}}{C}-COO^-$$

hydroxypyruvate

Intermediates, not in order:

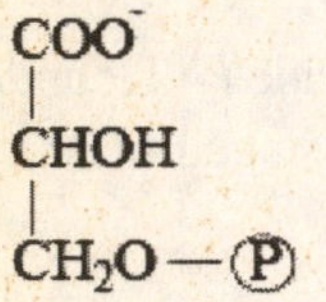

$$\begin{array}{c} COO^- \\ | \\ CHOH \\ | \\ CH_2O-\textcircled{P} \end{array} \qquad \begin{array}{c} COO^- \\ | \\ CHO-\textcircled{P} \\ | \\ CH_2OH \end{array} \qquad \begin{array}{c} COO^- \\ | \\ CHOH \\ | \\ CH_2OH \end{array} \qquad \begin{array}{c} COO^- \\ | \\ C-O-\textcircled{P} \\ \parallel \\ CH_2 \end{array}$$

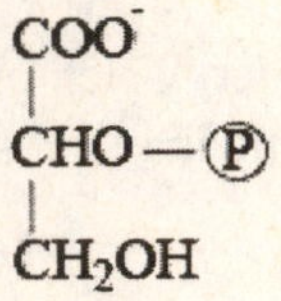

| A | B | C | D |

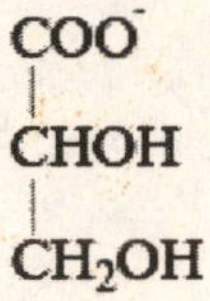

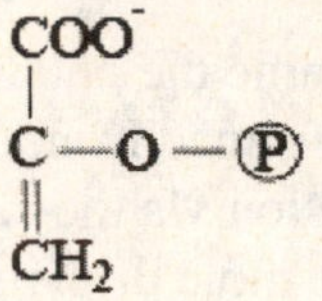

4. A hereditary disorder called encephalomyelopathy (Leigh's disease) results in accumulation of pyruvate and lactate in blood. This disorder is due to a deficiency of the enzyme pyruvate carboxylase. Speculate on the effect this deficiency might have on the capacity of the individual to perform continuous work or exercise.

5. The equilibrium constant for the formation of citrate catalyzed by citrate synthase is 5×10^5.

$$\text{oxaloacetate + acetyl CoA} \xrightarrow{\text{citrate synthase}} \text{citrate + CoASH}$$

(a) Calculate the change in standard free energy, $\Delta G^{\circ\prime}$, for this reaction at 25°C. Citrate can be transported from the mitochondrion into the cytosol and can be converted to acetyl CoA and oxaloacetate by the enzyme citrate lyase, according to this reaction:

$$\text{citrate + CoASH + ATP} \xrightarrow{\text{citrate lyase}} \text{oxaloacetate + acetyl CoA + ADP + P}_i$$

(b) Using information from text Table 10.1, calculate the change in standard free energy, $\Delta G^{\circ\prime}$, and the K_{eq} for this reaction at 25°C.

(c) Why is ATP required for this reaction? Is this requirement in harmony with the efficient operation of the citric acid cycle?

Electron Transport and
Oxidative Phosphorylation

CHAPTER

14

I. KEY TERMS

With the help of this study guide, your textbook, and class notes, you should be able to define and explain the significance of the following terms:

binding-change mechanism
chemiosmotic theory
coupled
oxidative phosphorylation
P:O ratio

proton motive force
Q cycle
respiratory electron transport chain
uncoupler

II. EXERCISES

A. True-False

_____ 1.   Both membranes of mitochondria have approximately the same composition.

_____ 2.   Cytochromes are iron-containing proteins that transfer electrons in the respiratory chain.

_____ 3.   In a normally functioning electron transport system, electron transport does not occur without ATP formation.

_____ 4.   Cytochrome *c* oxidase is the respiratory complex that directly transfers electrons to oxygen.

_____ 5.   All components of the respiratory chain are proteins.

_____ 6.   Each complex of the respiratory electron transport chain contains iron sulfur clusters that can perform one-electron transfers.

_____ 7.   Proton concentration increases in the mitochondrial matrix prior to ATP formation.

_____ 8.   In intact mitochondria, a respiratory inhibitor such as rotenone stops electron transport but has no effect on ATP production.

_____ 9.   The cytochrome *c* oxidase complex (Complex IV) utilizes copper ions as well as iron ions to achieve the transfer of electrons to oxygen.

_____ 10. 2,4-Dinitrophenol allows increased oxygen consumption even when mitochondria are deprived of ADP.

_____ 11. Oxidation of one mole of succinate by the electron transport system produces three moles of ATP.

_____ 12. The production of ATP requires energy that is supplied by a proton concentration gradient.

_____ 13. The inner membrane knobs contain an enzyme that, *in vitro*, was shown to catalyze the conversion of ATP to ADP and P_i.

_____ 14. Each of the four respiratory complexes translocates protons to the intermembrane space.

_____ 15. The synthesis of one ATP requires passage of four H^+ through the F_0 channel of Complex V into the mitochondrial matrix.

B. Short Answers

1. The heme-containing respiratory pigments are the _____.

2. Groups of solubilized proteins and other factors isolated from mitochondria that individually catalyze reactions from a segment of the respiratory chain are called _____.

3. The number of respiratory complexes in the electron transport chain is _____. These complexes contain multiple protein subunits and various other _____ that are directly involved in electron transfer.

4. The two mobile electron carriers in the electron transport system are _____ and _____.

5. The only electron-transferring component of the respiratory chain that is a lipid is _____.

6. Two inhibitors mentioned in your text that inhibit the transfer of electrons to oxygen by cytochrome *c* oxidase are _____ and _____.

7. The respiratory complex intimately connected with the citric acid cycle is _____. This complex contains _____, a citric acid cycle enzyme.

8. The machinery for ATP synthesis is located in the _____.

9. The theory proposed by _____ suggests that ATP formation is driven by the proton concentration gradient across the inner mitochondrial membrane. This hypothesis is called the _____.

10. The ratio of the moles of ATP produced per mole of oxygen atom reduced in the electron transport chain is called the _____.

11. The ADP/ATP carrier that exchanges mitochondrial ATP for/and cytosolic ADP is called _____.

C. Problems

1. How many protons are translocated by Complex I as NADH is oxidized and Q is reduced to QH_2?

2. Write the net reaction for the oxidation of NADH by the electron transport-oxidative phosphorylation process.

3. Write the net reaction for the oxidation of succinate by the electron transport-oxidative phosphorylation process.

4. Why does the oxidation of succinate lead to the formation of only about 1.5 ATP per mole rather than 2.5

ATP per mole as does NADH?

5. Using information in text Table 14.1, determine the free energy change for the transfer of a pair of electrons from NADH to the FMN of Complex I.

6. Explain how each of the following is affected when antimycin A is added to functioning mitochondria.

   (a) oxygen consumption   (b) ATP production
   (c) NAD^+ regeneration   (d) heat production

7. Explain the effects of 2,4-dinitrophenol on mitochondria with respect to the same four processes listed in the previous question.

8. In Section 10.9.B of your text, the standard free energy change for the transfer of electrons from NADH to oxygen was calculated to be -220 kJ/mol. Verify this value with the information given in text Table 14.2.

9. If ΔG = -19.4 kJ/mol for the entry of one H^+ into the mitochondrial matrix (see text equation 14.6) and four H^+ pass through the inner mitochondrial membrane per ATP formed, what percentage of the *available* ΔG is conserved by ATP formation?

10. What is the mechanism by which 2,4-dinitrophenol causes uncoupling of oxidative phosphorylation from the electron transport process?

11. What is the role of transport proteins in mitochondrial synthesis and cytoplasmic consumption of ATP?

12. How does oligomycin inhibit ATP synthesis?

13. ATP is produced in the mitochondrial matrix, but most ATP is used in the cytosol. Can ATP freely diffuse from the mitochondrial matrix into the cytosol?

D. Additional Problems

1. In the 1940s, 2,4-dinitrophenol was briefly used as an agent for weight reduction. Some individuals experienced elevated temperature, a marked increase in oxygen consumption, weakness, and profuse sweating. Some fatalities led to a ban of this agent for weight reduction. (a) Explain how this agent might achieve weight reduction. (b) Explain the symptoms observed in persons treated with 2,4-dinitrophenol.

2. Respiratory Complex IV, cytochrome *c* oxidase, catalyzes the transfer of electrons to oxygen. The proton concentration gradient is also affected by this transfer of electrons. "The effect is the same as a net transfer of four H^+ for each pair of electrons." (a) Write a reaction equation for the reduction of one mole of molecular oxygen by the action of Complex IV. (b) Explain the above statement in quotes.

3. Show that the transfer of electrons from NADH to FMN of Complex I, under standard conditions, does not provide enough energy to form ATP, but that the transfer of electrons from NADH to coenzyme Q does supply the required energy for one ATP to be formed.

4. Thermogenesis occurs in the brown adipose tissue of hibernating animals. What does the process do? What mechanism is involved and what stimulates the mechanism?

5. Imagine that you have isolated fragments of the mitochondrial inner membrane. You believe that these fragments contain one or more of the four respiratory complexes. By additional experimentation, you find that neither succinate nor NADH is oxidized, in the presence of Q, by the isolated fragments but that the following reaction is catalyzed:

$$2 \text{ Cyt } c(Fe^{2+}) + 1/2 \text{ } O_2 + 2 \text{ H+} \rightarrow 2 \text{ Cyt } c(Fe^{3+}) + H_2O$$

What respiratory complex (or complexes) is (are) in your isolated fragments?

Photosynthesis

I. KEY TERMS

With the help of this study guide, your textbook, and class notes, you should be able to define and explain the significance of the following terms:

accessory pigment
amyloplasts
C₄ pathway
Calvin cycle
chlorophyll
chloroplast
Crassulacean acid metabolism
cyclic electron transport
dark reactions

grana
light reactions
light-harvesting complex
lumen
photon
photophosphorylation
photorespiration
photosynthesis
photosystem

reaction center
reductive pentose phosphate
    cycle
Rubisco
special pair
stomata
stroma
thylakoid membrane
Z-scheme

II. EXERCISES

A. True-False

_____ 1. Photosynthetic eukaryotes usually have both mitochondria and chloroplasts that contribute to energy production.

_____ 2. During photosynthesis, water is oxidized even though the process is thermodynamically unfavorable.

_____ 3. The light reactions of photosynthesis occur only in daylight and the dark reactions of photosynthesis occur only at night.

_____ 4. Oxygen is produced as a consequence of the dark reactions of photosynthesis.

_____ 5. The oxygen produced by green plants comes directly from the oxygen atoms of CO_2 taken in by the plant.

_____ 6. Although chlorophyll *a* and chlorophyll *b* have specific absorption maxima, plants are able to absorb and use energy across the visible spectrum.

_____ 7. All photosynthetic organisms rely on either chlorophyll *a* or chlorophyll *b* to absorb light.

_____ 8. A proton motive force is created in chloroplasts as protons are pumped across the thylakoid membrane into the stroma.

_____ 9. The functional units of photosynthesis in plants are the photosystems, which are complexes of pigment molecules and proteins located in the thylakoid membrane.

_____ 10. The light reactions produce one ATP and one NADPH for each pair of electrons transferred from water, but, in the dark reactions, two molecules of NADPH and three molecules of ATP are consumed in the reduction of one molecule of CO_2 to carbohydrate.

_____ 11. In plants, sucrose and starch are both formed in the cytosol.

B. Short Answers

1. The two major processes of photosynthesis are the _____ reactions that produce _____, _____, and _____, and the _____ reactions that use CO_2, ATP, and NADPH to produce _____.

2. In eukaryotic organisms, photosynthesis occurs in organelles called _____.

3. Within a chloroplast, the _____ membrane is arranged in stacks called _____.

4. Protein-bound pigment molecules that harvest light for the photosystems in the thylakoid membrane are called _____.

5. The process by which ATP is formed as photosynthesis occurs is called _____.

6. Each photosystem contains a _____, which is a complex of proteins, electron-transporting cofactors, and two chlorophyll molecules called the _____.

7. In a chloroplast, the light-dependent reactions take place in the _____.

8. The most powerful oxidizing agent in biochemical reactions is _____.

9. An iron-sulfur protein coenzyme involved in the transfer of electrons that achieves the reduction of $NADP^+$ to NADPH is _____.

10. The _____ is associated with photosystem II on the lumen side of the granal lamella and is directly involved with the production of oxygen from water.

11. In chloroplasts, the reductive conversion of CO_2 into carbohydrates occurs in the _____.

12. The compound formed when CO_2 is fixed into an organic product in the RPP cycle is _____.

13. The cofactor required for the reduction of 1,3-*bis*phosphoglycerate to glyceraldehyde 3-phosphate is _____.

14. The enzyme that catalyzes the formation of 3-phosphoglycerate from CO_2 and ribulose 1,5-*bis*phosphate is called _____, also known as _____.

15. The four photosynthetic complexes that operate in a chloroplast are _____, _____, _____, and _____.

C. Problems

1. What is the function of the phytol side chain of the various chlorophyll molecules?

2. What are the three states of a chlorophyll molecule with respect to the absorption and utilization of light energy?

3. Summarize the events that occur in each of the two major processes of photosynthesis.

4. Calculate the number of moles of water that must be oxidized in photosynthesis to provide enough reducing power, as NADPH, to form one mole of sucrose.

5. Prove that light of shorter wavelengths provides a greater amount of energy for photosynthesis than does light of longer wavelengths. Compare the energy of a mole of photons of light of 700 nm with that of light of 400 nm. Recall that the energy of a photon is $E = h\nu$, where h is Planck's constant (6.626×10^{-34} J s) and ν is the frequency of the light, related to the speed of light and its wavelength, λ, by $c = \lambda\nu$. The speed of light, c, is 3.00×10^7 nm/s.

6. Explain the role of carotenoids in photosynthesis.

7. Indicate at least three ways in which photosynthesis is similar to electron transport and oxidative phosphorylation.

8. Emerson and Arnold found that oxygen yields of *Chlorella* (a green algae) subjected to flashes of light leveled off at light intensities that were sufficient to excite only a small fraction of 2400 chlorophyll molecules. Additionally, they found that eight photons of energy must be sequentially absorbed to produce one oxygen molecule. They described a functional cluster as the number of chlorophyll molecules that utilize one photon of energy. (a) How many chlorophyll molecules comprise a functional cluster? (b) How many functional clusters are needed to produce one oxygen molecule? (Note: This work gave rise to the concept of a photosynthetic unit or a photosystem.)

9. The amount of ATP produced in the light-dependent reactions of photosynthesis does not seem adequate to fill the ATP requirements of the dark (light-independent) reactions. How are the levels of NADPH and ATP balanced by the light reactions at the relative levels of their use in CO_2 fixation?

10. What is different about the locations of photosystem I, PSI, and photosystem II, PSII? What is the functional importance of this difference?

11. In 1965, Jagendorf and Uribe studied ATP formation in spinach chloroplasts. In one study, they soaked the chloroplasts in pH 4.0 buffer, which allowed the pH to drop to 4.0 inside the chloroplasts; then they quickly changed the external pH to 8.0. This resulted in the formation of ATP from ADP and P_i. How did this study contribute to our present understanding of photosynthesis?

12. (a) How many molecules of CO_2 are required for photosynthetic hexose formation? (b) Explain how the CO_2 molecules are utilized.

13. What is the function of 2-carboxyarabinitol 1-phosphate in plants?

14. Plants such as the orchid, cactus, and pineapple keep their stomata closed during the day to minimize water loss. How do these plants, with closed stomata, get enough CO_2 for carbon assimilation?

15. Where is starch synthesized? What is the energy involved in its synthesis?

D. Additional Problems

1. In a photosystem, both the antenna complex and the reaction center contain chlorophyll molecules. Only the special pair chlorophyll molecules in the reaction center are intimately involved in the conversion of light energy to chemical energy. Speculate on the difference in the chlorophylls in the reaction center from those of the antenna complex that allows them to perform different functions. (Note: You may need to consult sources other than your text to answer this question.)

2. Figure 15.1 shows a plot of the 1966 data of Jagendorf and Uribe in which the ATP yield of spinach chloroplasts buffered at pH 4.0 was determined when their external (second stage) pH was increased. The chloroplasts had previously been illuminated while buffered at pH 4.0. How does this yield of ATP per chlorophyll compare with the findings of Emerson and Arnold discussed earlier (Problem C.8.)?

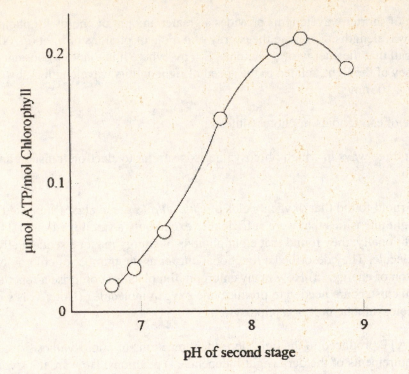

Figure 15.1   ATP Production of Spinach Chloroplasts Subjected to Different pH Gradients

3. The enzyme Rubisco accepts both CO_2 ($K_m = 20$ µM, $k_{cat} = 3$ s^{-1}) and O_2 ($K_m = 200$ µM) as substrates, which compete for the same active site. (a) What are the normal relative rates of the two processes *in vivo*? (b) Are there circumstances under which the competing reactions can limit the growth of the plant? (c) Photorespiration in plants with the C_4 pathway (corn, sorghum, sugarcane) is minimal and does not limit growth of these plants. By what process is photorespiration minimized in these plants? (Note: You may have to consult sources other than your text to answer all parts of this question.)

4. If solar energy of about 7 J (cm^2 s)$^{-1}$ is absorbed by a 10 cm^2 area of a leaf and results in the production of 5.53 grams of glucose in one hour, what is the percent efficiency of the hexose production process? For the *in vitro* oxidation of glucose, $\Delta G^{\circ\prime} = -686$ kcal/mol or 2870 kJ/mol.

5. The following equation has been used to calculate the free energy change for the transfer of protons across the inner mitochondrial membrane to the intermembrane space.

$$\Delta G = 2.3 \, RT \, [pH(in) - pH(out)] + ZF\Delta\Psi,$$

where R is the gas constant, T is the absolute temperature, Z is the charge on the proton, F is the Faraday, and $\Delta\Psi$ is the membrane potential. Assume a temperature of 25 °C, [ATP]/[ADP][P$_i$] = 1000, and determine the pH difference across the thylakoid membrane necessary to drive the formation of ATP during photosynthesis. Assume that a transfer of three protons is involved and that the membrane is freely permeable to Ca^{2+} and Cl^- ions.

6. In 1945, Melvin Calvin and coworkers showed that 3-phosphoglycerate was the first compound to show a

^{14}C-label when $^{14}CO_2$ was used in their studies of photosynthesis. (a) What did this result suggest about the size of the carbon chain of the CO_2 acceptor? (b) How was this conclusion modified when it was discovered that there were two moles of 3-phosphoglycerate formed for each mole of CO_2 fixed? (c) Where is the label located in 3-phosphoglycerate?

Lipid Metabolism

I. KEY TERMS

With the help of this study guide, your textbook, and class notes, you should be able to define and explain the significance of the following terms:

acidic phospholipid
acyl carrier protein
carnitine
eicosanoid

isopentenyl pyrophosphate
ketogenesis
lipoprotein
neutral phospholipid

triacylglycerol
β-oxidation pathway

II. EXERCISES

A. True-False

_____ 1. L-Carnitine is an intermediate in the β-oxidation pathway.

_____ 2. Most naturally occurring fatty acids have even-numbered carbon chains.

_____ 3. Fatty acids are degraded two carbons at a time from the carboxyl end of the molecule.

_____ 4. β-oxidation of eicosanoyl CoA yields 10 NADH, 10 QH_2, and 10 acetyl CoA.

_____ 5. The activation of fatty acids required for β-oxidation is an energy-consuming process.

_____ 6. The mitochondrial matrix is the site of both the degradation and synthesis of fatty acids.

_____ 7. The liver is the primary site of ketone body formation.

_____ 8. Almost all of the NADPH required for the synthesis of fatty acids is furnished by the pentose phosphate pathway.

_____ 9. The carbon atoms of the steroid ring structure of cholesterol are all derived from acetyl CoA.

_____ 10. A high ratio of LDL to HDL in the blood is associated with increased risks of cardiovascular disease.

_____ 11. Saturated fatty acids yield more ATP per gram than does glucose.

B. Short Answers

1. In mammals, the two major forms of stored energy are _____ and _____.

2. The products of β-oxidation of fatty acids are _____, _____, and _____.

3. Fatty acyl CoA molecules are oxidized by a repeated series of four enzyme-catalyzed steps that are _____, _____, _____, and _____.

4. In β-oxidation, the enzyme that catalyzes cleavage of its substrate to produce acetyl CoA as a product is _____.

5. The molecules known as ketone bodies are _____, _____, and _____.

6. The eukaryotic organelles in which β-oxidation occurs are _____ and _____.

7. Acetyl CoA is required for fatty acid synthesis, but the form through which carbons are supplied for extending fatty acid carbon chains is _____.

8. The key regulatory enzyme of fatty acid synthesis is _____.

9. Enzymes that create double bonds in saturated fatty acyl CoA molecules are called _____.

10. Short-lived regulatory molecules, called local regulators, include prostacyclin, thromboxane A_2, and other prostaglandins that are derived from prostaglandin H_2, an eicosanoid derived from _____.

11. In addition to the prostaglandins and thromboxanes formed as one class of eicosanoids, another class includes the _____, molecules involved in allergic response.

12. _____ is the name of a drug used to lower blood cholesterol levels. It is a competitive inhibitor of HMG-CoA reductase, the enzyme that catalyzes the first committed step in cholesterol biosynthesis.

13. Noncyclic precursors of cholesterol that have 10-, 15-, and 30-carbon chains, respectively, are _____, _____, and _____.

14. In eukaryotic cells, biosynthesis of lipids including phosphatidylcholine, phosphatidylserine, and phosphatidylinositol in eukaryotic cells occurs in the _____.

C. Problems

1. What are the typical chain lengths for fatty acids synthesized in the cell? What is the reason for this limit?

2. Write the overall (net) equation for the synthesis of stearate.

3. Write the overall (net) equation for the β-oxidation of stearoyl CoA.

4. The activation of a fatty acid molecule is said to consume two ATP equivalents. Why is that, since only one ATP molecule is used?

5. Why are the natural unsaturated fatty acids not suitable substrates for the enzymes of the β-oxidation pathway when one intermediate of the pathway is unsaturated?

6. Under what conditions do liver mitochondria synthesize acetoacetate and β-hydroxybutyrate from some of the acetyl CoA formed from fatty acids?

7. If ^{14}C-labeled bicarbonate is used in cellular fatty acid biosynthesis, where does the label appear after a fatty acid like palmitate has been synthesized?

8. You learned previously that lineolate is an essential fatty acid. Now, from information in this chapter of your text, you can explain why—please do.

9. Cholestyramine is a nondigestible anion-exchange resin that has been taken orally to help lower human cholesterol levels. How does it cause blood cholesterol levels to decrease?

10. Your text shows that, compared with the ATP yield of glucose, palmitate yields about 24% more ATP on a per carbon basis. Is this amount of extra ATP yield the same for all saturated fatty acids? To determine the answer, calculate the ATP yield for the oxidation of stearic acid and compare it with that of glucose, on a per carbon basis.

11. (a) What are the functions of the citrate transport system? (b) Is energy required for or produced by this system? (You may have to consult the previous edition of your text or another source to answer this question.)

12. What is the key regulatory enzyme in fatty acid synthesis and what reaction does it catalyze?

13. You have two containers whose labels are missing. You know that one solution contains cholesterol and the other contains sphingomyelin. Suggest a simple laboratory procedure that would allow you to identify the solution that contained cholesterol.

14. How does aspirin alleviate pain and inflammation?

15. What is colipase? What is its function?

D. Additional Problems

1. Is there any biochemical basis for the statement that going on a carbohydrate-free diet can lead to a life-threatening situation?

2. Release of fatty acids for energy production is inhibited by insulin. In diabetics who lack insulin, massive amounts of fatty acids are released from adipose tissue. This and other factors lead to an acceleration of fatty acid oxidation. Can fatty acid oxidation proceed indefinitely? What are the consequences?

3. Why are lovastatin and cholestyramine ineffective in lowering cholesterol levels in individuals with familial hypercholesterolemia (FH)? (In individuals with FH, LDL-receptor proteins are produced at lower than normal levels, or not at all, in those individuals severely afflicted.)

4. Early studies showed that the protein avidin strongly inhibited the incorporation of acetyl CoA into fatty acid. What is the nature of this interference?

5. As a general rule, organisms without a glyoxylate cycle cannot convert saturated fatty acids to carbohydrates. What is a minor exception to this rule and why is it a *minor* exception?

6. Cyclooxygenase (COX) has two known isozyme forms, COX-1 and COX-2. Aspirin inhibits both of these forms. Why/how? More recently, research efforts were focused on development of aspirin substitutes that selectively inhibit COX-2. Why?

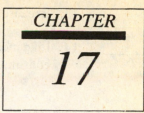

Amino Acid Metabolism

I. KEY TERMS

With the help of this study guide, your textbook, and class notes, you should be able to define and explain the significance of the following terms:

caspase	ketogenic	proteasome
essential amino acid	nitrogen cycle	turnover
glucogenic	nitrogen fixation	urea cycle
glucose-alanine cycle	nonessential amino acid	

II. EXERCISES

A. True-False

_____ 1.   Excess nitrogen is eliminated from birds, aquatic animals, and terrestrial vertebrates as urea.

_____ 2.   In mammals, only about one-fourth of the amino acids used for protein synthesis are essential (must be obtained in the diet).

_____ 3.   Amino acids derived from ingested proteins in the diet of mammals are used exclusively for cellular protein synthesis.

_____ 4.   The majority of transaminases exhibit a specificity for L-glutamate and α-ketoglutarate.

_____ 5.   Urea is synthesized in mammalian kidneys.

_____ 6.   Nitrogen fixation, catalyzed by the enzyme nitrogenase, is a process that requires energy.

_____ 7.   Nitrogen fixation via nitrogenase produces hydrogen gas as well as ammonia.

_____ 8.   Glutamine is a nitrogen donor in several biosynthetic reactions and represents a means of carrying nitrogen between tissues.

_____ 9.   Each of the 20 common amino acids can be described as being either glucogenic or ketogenic.

_____ 10. Arginase is an enzyme found at high activity levels in nearly all mammalian organs and tissues.

_____ 11. Protons from acids generated during metabolism are neutralized by bicarbonate in the blood. Bicarbonate must be regenerated, however, so that blood pH is maintained by the bicarbonate buffer system.

_____ 12. Nitroglycerin exerts its effect, dilation of coronary arteries, by virtue of being metabolized to nitric oxide.

_____ 13. Because of the importance of their functions, regulatory proteins have long half-lives.

_____ 14. Like most catabolic processes, the protein hydrolysis that occurs during turnover is not energy consuming.

B. Short Answers

1. A coenzyme required by the transaminases is _____. Its function in transamination is _____.

2. While most of the common amino acids can be formed by transamination, little or none of the two amino acids _____ and _____ is formed by transamination.

3. Amino acids whose carbon chains are catabolized to form acetyl CoA are said to be _____. Those that form intermediates of glycolysis or the citric acid cycle are said to be _____.

4. A mutation in the gene that codes for phenylalanine hydroxylase makes afflicted individuals unable to tolerate normal quantities of phenylalanine. The resulting disease is known as _____.

5. In the biosphere, the two major sources of nitrogen ultimately used for amino acid synthesis are _____ and _____.

6. The sources of the two nitrogen atoms and the carbon atom of urea are _____, _____, and _____, respectively.

7. One of the nitrogen atoms and the carbon atom of urea enter the urea cycle as a single molecule, namely _____.

8. The urea cycle requires energy to form urea. The equivalent of _____ ATP molecules is required.

9. The exchange of glucose and alanine between muscle and liver is called the _____.

10. As serine is catabolized, it is converted to glycine, which is broken down by the action of _____.

11. Several molecules of the protein _____ are attached to proteins targeted for degradation.

12. The multiprotein complex that hydrolyzes targeted proteins to peptides during the turnover process is called the _____.

13. With respect to amino acid synthesis, aspartate serves as precursor for the amino acids _____, _____, and _____.

C. Problems

1. What are the two major stages by which amino acids are degraded?

2. In the glucose-alanine cycle, alanine formed in muscle is transported by the blood to the liver where it is acted on by alanine transaminase. Using appropriate structures, write this transamination reaction equation.

3. Complete these word equations (no structures required).

   a. oxaloacetate + glutamate $\xrightarrow{\text{aspartate transaminase}}$

b. glutamine + aspartate $\xrightarrow{\text{asparagine synthetase}}$

4. To regenerate α-ketoglutarate required by transaminases, L-glutamate must be oxidatively deaminated in a reaction catalyzed by glutamate dehydrogenase. How can this deamination occur when the $\Delta G^{\circ\prime}$ for the reaction is +27 kJ/mol?

5. The answer to problem B.8. of the previous section indicated that the equivalent of four ATP molecules is required for the synthesis of one molecule of urea. Detail this energy expenditure.

6. Discuss the regulation of the urea cycle.

7. Explain/criticize the statement that glycine synthesis is dependent upon levels of folate and serine.

8. In text Section 17.3.D., you are told that proline is formed as a result of nonenzymatic cyclization of glutamate γ-semialdehyde to form a Schiff base that is then reduced. Using appropriate structures, write the reaction sequence for the formation of proline from glutamate γ-semialdehyde. You may use the appropriate number of electrons (as H atoms) rather than some specific reducing agent.

9. List seven compounds of central metabolism into which carbon atoms from the carbon chains of amino acids are channeled.

10. Separately list those amino acids that are ketogenic, those that are glucogenic and those that fall into both categories.

11. Explain how the synthesis of two nonessential amino acids is dependent upon two essential amino acids.

12. What are caspases?

13. List the compounds that supply the carbon and nitrogen atoms in the complex synthesis of histidine.

14. Complete the net word equation for the removal of the –NH₂ group from alanine, a process that involves two separate reactions.

D. Additional Problems

1. Imagine that you worked with Sir Hans Krebs during the time he was elucidating the urea cycle. You used slices of rat liver tissue and exposed these tissues to a solution containing urease. The gas evolved was collected and found to be 2.4 mL at STP conditions. How many moles of urea were present in the tissue slices? (Assume that none of the gas produced dissolved in the solution.)

2. Compare the function of the alanine-glucose cycle with that of the Cori cycle. Under what circumstances might each operate?

3. The ammonium ion is unable to enter the brain, but several known defects in urea cycle enzymes are known to lead to mental retardation. What is the connection?

4. The first nitrogen atom of the urea molecule enters the urea cycle as carbamoyl phosphate, formed from bicarbonate and ammonium ions. Detail the entry of the second nitrogen atom of the urea molecule and how the nitrogen carrier itself obtained the nitrogen.

5. When ^{18}O-labeled citrulline (amide oxygen) reacts with aspartate to form argininosuccinate, as catalyzed by argininosuccinate synthetase, (a) where does the label appear? (b) What does this suggest about a possible intermediate of the reaction? (Note: Sources other than your text may be required to answer this question.)

CHAPTER

18

Nucleotide Metabolism

I. KEY TERMS

With the help of this study guide, your textbook, and class notes, you should be able to define and explain the significance of the following terms:

5-phosphoribosyl 1-
  pyrophosphate

inosine 5'-monophosphate
purine nucleotide cycle

salvage

II. EXERCISES

A. True-False

_____ 1. The nitrogenous bases of nucleotides and nucleosides are of two types—the purines, which are heterocycles containing six atoms, and the pyrimidines, which are heterobicycles containing nine atoms.

_____ 2. Nucleotide biosynthesis occurs in virtually all tissues.

_____ 3. To synthesize the purine ring, liver cells require metabolic sources of glycine, formate, glutamine, CO_2, and aspartate.

_____ 4. Tissues synthesize the purine ring and then attach it to ribose 5-phosphate to make nucleotides.

_____ 5. Inosine 5'-monophosphate (IMP) is an intermediate from which both 5'-AMP and 5'-GMP are synthesized.

_____ 6. GTP is required for the synthesis of AMP (and subsequently ATP), and ATP is required for the synthesis of GMP (and subsequently GTP).

_____ 7. To synthesize the pyrimidine ring, liver cells require aspartate, glutamine and bicarbonate.

_____ 8. The first step in the pyrimidine ring biosynthesis involves formation of carbamoyl phosphate, catalyzed by carbamoyl phosphate synthetase. This is the same enzyme that operates in the urea cycle.

_____ 9. Uric acid, an avian excretion product, is a pyrimidine.

_____ 10. Orotate is the end product of the pyrimidine biosynthetic pathway and is the derivative from which the pyrimidine nucleosides and nucleotides are made.

_____ 11. Like the purine ring system, the pyrimidine ring system is formed using ribose 5-phosphate as the foundation upon which it is built.

_____ 12. dTMP is formed by the reduction of the ribose moiety of ribothymidylate.

_____ **13.** Most free purine and pyrimidine molecules are salvaged, but some are catabolized. Catabolism of both purine and pyrimidine molecules leads to the excretory product uric acid.

B. Short Answers

1. The components of a nucleotide are a/an _____, a/an _____, and _____.

2. The active form of ribose 5-phosphate required for nucleotide biosynthesis is _____.

3. In liver cells, the location of purine biosynthesis is _____.

4. Of the two fused rings of purine nucleotides, the one formed first in the cell is _____.

5. Four feedback inhibitors of enzymes of the purine biosynthetic pathway are _____, _____, _____, and _____.

6. Carbamoyl phosphate is involved in the biosynthesis of which type of nucleotides, the pyrimidines or the purines? _____

7. The recycling of products from the normal cellular breakdown of nucleic acids is an energy-conserving process called _____.

8. _____ is the product of purine nucleotide catabolism that is involved in gout.

9. CTP is formed from UTP by transformation of the keto group at C-4 of the pyrimidine ring to an amino group. The source of the amino group is _____.

10. Two enzymes, sometimes called salvage enzymes, which are responsible for recycling the bases adenine, hypoxanthine, and guanine to the corresponding nucleotides AMP, IMP, and GMP are _____ and _____.

11. A pathway in muscle that produces ammonia and fumarate under exercise/work conditions and involves nucleotides is called the _____.

12. Two thioesters of central metabolism formed from the catabolism of pyrimidines are _____ and _____.

13. The end product of purine catabolism in most animals (excluding humans) is _____.

C. Problems

1. List the products, and the number of moles of each, formed by the hydrolysis of (a) cytidine, (b) inosine monophosphate, and (c) dGTP.

2. Glutamine is required in the biosynthetic pathways of both the purines and the pyrimidines. What is its involvement in purine biosynthesis?

3. Since PRPP is a donor of ribose 5-phosphate in more than a dozen reactions, it might seem that control of its synthesis by control of PRPP synthetase would be very likely. Is this a major control site?

4. In general, biosynthesis is energy-consuming. Starting with ribose 5-phosphate, how many ATP equivalents are consumed in the synthesis of 5'-AMP? Assume sufficient glutamine and other required

materials are available. Also ignore any reduced cofactors formed that could provide ATP upon reoxidation.

5. Glutamine-PRPP amidotransferase catalyzes the first committed step of purine nucleotide synthesis. It is inhibited by both guanosine and adenosine phosphates (feedback inhibition). Discuss the likelihood of (a) more than one inhibitor-binding site on the enzyme, and (b) complete inhibition by any one of the inhibitors alone.

6. Write the overall (net) equation for the cellular formation of UMP. (Structures are not required.)

7. What is the fate of the C-1 carboxyl group of aspartate in the biosynthesis of UMP?

8. Step 6 in the biosynthesis of IMP involves a carboxylation catalyzed by aminoimidazole ribonucleotide carboxylase (AIR carboxylase). What is unusual about the requirements of this enzyme?

9. Using appropriate structures, write the equation for the formation of CTP from UTP.

10. If 3-[^{14}C]-serine is supplied to cells performing nucleotide biosynthesis, where will the label appear in dTMP?

11. Explain how deoxyribonucleotides are synthesized from ribonucleotides.

12. (a) What hereditary deficiency leads to Lesch-Nyhan syndrome? (b) Why is the disease usually restricted to males? (c) What are the clinical symptoms of the disease? (d) Why do afflicted individuals have greatly increased *de novo* synthesis of purine nucleotides?

D. Additional Problems

1. Carbamoyl phosphate I uses ammonia as the amide nitrogen source of carbamoyl phosphate used in the urea cycle. Carbamoyl phosphate II uses glutamine as source of the amide nitrogen used in pyrimidine synthesis. Why does allosteric control of pyrimidine synthesis via carbamoyl phosphate II not interfere with urea synthesis?

2. Assume that ATP, glutamine, and ^{14}C-HCO$_3^-$ are incubated with tissue slices that contain the enzymes and other materials necessary for pyrimidine nucleotide biosynthesis. If orotidine 5'-phosphate (OMP), UMP, and CO$_2$ are isolated from the system, in which of these should the label be found?

3. The purine nucleotide cycle provides the citric acid cycle intermediate fumarate to assist operation of the citric acid cycle and production of ATP for energy needs. NH$_4^+$ is also produced by the purine nucleotide cycle. If all NH$_4^+$ so produced is eliminated as urea, what is the energy cost per mole of fumarate produced by the purine nucleotide cycle?

4. What would be the consequences of a hereditary deficiency of the enzyme deoxyuridine triphosphate diphosphohydrolase (dUTPase)?

5. von Gierke's glycogen storage disease is caused by a deficiency of glucose 6-phosphatase. This can lead to overproduction of uric acid and to gout. Explain the connections.

6. Azaserine is a glutamine analog that has been used as an affinity label. (a) What nucleotide biosynthetic enzymes would be affected by administration of azaserine or similar antibiotics? (b) What is the mechanism of such inhibition?

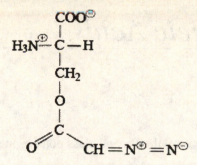

Azaserine

7. Explain how 5-fluorouracil combats some types of cancer.

CHAPTER

19

Nucleic Acids

I. KEY TERMS

With the help of this study guide, your textbook, and class notes, you should be able to define and explain the significance of the following terms:

A-DNA	melting temperature (T_m)	ribonuclease
antiparallel	messenger RNA	ribonucleotide
base pair	minor groove	ribosomal RNA
B-DNA	nuclease	rise
chromatin	nucleic acid	small RNA
denaturation	nuclein	solenoid
double helix	nucleoside	stacking interactions
endonuclease	nucleosome	sticky end
exonuclease	nucleotide	supercoiling
genome	palindrome	topoisomerase
histone	pitch	transcription
kilobase pair	restriction endonuclease	transfer RNA
major groove	restriction map	Z-DNA
melting curve	restriction methylase	

II. EXERCISES

A. True-False

_____ 1. Nucleotides are molecules that contain only a sugar and a nitrogenous base.

_____ 2. Nucleosides are obtained as breakdown products of nucleic acids and were found to contain a pentose and a nitrogenous base held to the pentose by a β-N-glycosidic bond.

_____ 3. Uracil and thymine are members of the pyrimidine family and are typically found as components of DNA.

_____ 4. The nucleoside moieties of nucleic acids exist predominately in the *anti* conformations.

_____ 5. Nuclein was the term used by Miescher for the mixture of nucleic acids and protein isolated from white blood cells.

_____ 6. Chargaff determined that the molar ratios of adenine to thymine and of guanine to cytosine in DNA are about one. This means that the sum of the adenine + thymine bases will be equal to the sum of the guanine + cytosine bases.

_____ 7. The Watson-Crick model of DNA is a right-handed helix of two separate chains, both oriented in the same direction.

_____ 8. The ultraviolet absorbance of one mole of duplex (double-stranded) DNA is greater than that of the same amount of denatured DNA.

_____ 9. Endonucleases are enzymes that catalyze the cleavage of phosphodiester linkages in the interior of nucleic acids.

_____ 10. Some eukaryotic chromosomes are circular.

_____ 11. DNA can be hydrolyzed in the laboratory by dilute NaOH.

_____ 12. Hydrolysis of RNA by the catalytic action of pancreatic ribonuclease or by dilute NaOH involves formation of a 2',3'-cyclic nucleoside monophosphate intermediate.

_____ 13. In many bacteria, there are specific restriction methylases and restriction endonucleases that recognize the same sequence of bases in the substrate DNA.

_____ 14. Sticky ends are pieces of single-stranded DNA formed by the action of certain endonucleases as a result of duplex DNA denaturation.

_____ 15. In eukaryotes, the genome consists of nuclear DNA as well as the mitochondrial and chloroplast DNA.

_____ 16. In all species, the ratio of purine bases to pyrimidine bases in DNA is always 1:1.

_____ 17. Under physiological conditions, nucleic acids are polyanions.

_____ 18. A-DNA, B-DNA, and Z-DNA all exist *in vivo*.

B. Short Answers

1. 2,4-Dioxopyrimidine is more commonly known as _____.

2. 2-Oxo-4-aminopyrimidine is more commonly known as _____.

3. Deoxythymidine-5'-triphosphate is commonly abbreviated as _____.

4. DNA polymerase requires _____ as substrates for the extension of a DNA chain.

5. The hydrated, more common form of DNA is _____.

6. The type of DNA that exists as a left-handed helix is _____.

7. The allowed base pairing in DNA is thymine with _____ and cytosine with _____.

8. The process of using heat to dissociate duplex DNA into separate single strands is called _____. The temperature at which half of the DNA has become single stranded is called the _____.

9. Bacterial enzymes that recognize and catalyze the cleavage of foreign DNA are called _____.

10. The DNA sequences recognized by most restriction endonucleases are called _____.

11. Eukaryotic DNA has a number of basic proteins called _____ associated with it. The resulting nucleoprotein complex is called a/an _____.

12. Four types of interactions that influence the stability of duplex DNA are _____ , _____ , _____ , and _____ .

13. Enzymes that break DNA, unwind or overwind the double helix, and rejoin the strands are called _____ .

14. RNA molecules that carry information from DNA to the ribosomes for protein synthesis are called _____ .

15. When certain bases in DNA are methylated for protection from the cell's own restriction endonucleases, the
–CH_3 group is supplied by _____ .

16. The basic proteins called histones that associate with DNA are found only in _____ cells.

C. Problems

1. In what state of ionization does ATP normally exist in cells?

2. The guanine content of a double-stranded DNA sample was determined to be 17% (of total bases in the sample). Name the other bases present in the DNA and the corresponding percentages of each.

3. In a B-DNA segment that is about 51 nm in length, what is the total number of bases?

4. Alexander Rich synthesized hexadeoxynucleotide chains with repeating cytosine and guanine bases. The association of two such strands forms a double helix with the Z-DNA conformation. These two strands are said to be self-complementary. Illustrate this self-complementary aspect.

5. Of the total bases present, what is the approximate mole% of adenine in a DNA sample with a T_m of 90°C, based on the data presented in Figure 19.1?

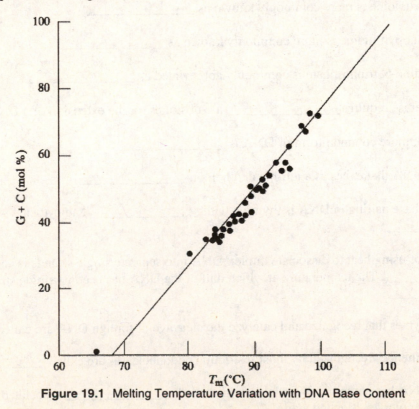

Figure 19.1 Melting Temperature Variation with DNA Base Content

6. What is the expected T_m for a DNA sample with a thymine content of 20%?

7. Explain the mechanism by which some bacteria recognize and destroy foreign DNA without destroying their own DNA.

8. Why does duplex DNA form a helix rather than existing as two side-by-side antiparallel strands?

9. What is the most abundant type of RNA in cells and what is its function?

10. What is the composition of a nucleosome and what is its relationship to the 30-nm fiber?

11. Consider the restriction endonucleases in text Table 19.4. Which enzymes do not produce sticky ends when cleaving DNA?

D. Additional Problems

1. If the largest human chromosome contains 2.4×10^8 base pairs, (a) how many nucleosomes does it contain? (b) If this chromosome contained 2000 loops of the 30-nm fiber as it is attached to an RNA-protein scaffold, how many nucleosomes are in each loop, on the average?

2. You are a biochemist studying DNA isolated from a particular type of phage. You note that the DNA serves as a substrate for DNase I and DNase II, but not for snake venom phosphodiesterase or spleen phosphodiesterase. Further, analysis of base content indicates that the ratio of thymine to adenine is 1.36. Explain these data. (Table 19.1 contains information that may be helpful in the solution of this problem.)

Table 19.1 Some Representative Nucleases

	Substrate	Product
Exonucleases (3'-5')		
Exonuclease I (*E. coli*)	Single-stranded DNA	Nucleoside 5'-phosphates
DNA polymerase (*E. coli*)	Single-stranded DNA*	Nucleoside 5'-phosphates
Exonuclease III (*E. coli*)	Double-stranded DNA*	Nucleoside 5'-phosphates
Bacillus subtilis exonuclease	Double-stranded DNA, RNA	Nucleoside 5'-phosphates, oligonucleotides
Snake venom phosphodiesterase	Single-stranded DNA or RNA	Nucleoside 5'-phosphates (no base specificity)
Exonucleases (5'-3')		
Bacillus subtilis exonuclease	Single-stranded DNA	Nucleoside 3'-phosphates (also works from 3' end)
Exonuclease VII (*E. coli*)	Single-stranded DNA	Oligonucleotides (also works from 3' end)
Neurospora crassa exonuclease	Single-stranded DNA, RNA	Nucleoside 5'-phosphates
Exonuclease VI or DNA polymerase (*E. coli*)	Double-stranded DNA	Nucleoside 5'-phosphates
Bovine spleen phosphodiesterase	Single-stranded DNA or RNA	Nucleoside 3'-phosphates (no base specificity)
Endonucleases		
Neurospora crassa endonuclease	Single-stranded DNA, RNA	Polynucleosides with 5'-phosphate termini
Endonuclease I (*E. coli*)	Single- or double-stranded DNA	Polynucleosides with 5'-phosphate termini
DNase I (bovine pancreas)	Single- or double-stranded DNA	Polynucleosides with 5'-phosphate termini
DNase II (calf thymus)	Single- or double-stranded DNA	Polynucleosides with 3'-phosphate termini
RNase A (bovine pancreas)	RNA	Polynucleosides with 3'-phosphate termini

*Also catalyzes hydrolysis of nicked and gapped double-strand DNA.

3. In 1963, Jerome Vinograd studied circular DNA from polyoma virus. The DNA had the expected 1:1 ratio for A-T and G-C pairs, but two DNA bands were observed when the DNA was centrifuged. Since both types of DNA were circular and of identical base composition, why were there two bands?

4. Pancreatic DNase I catalyzes the internal cleavage of single-stranded or duplex B-DNA somewhat at random and does not cleave Z-DNA at all. Use this information to draw some conclusions about the manner in which DNase I functions. (Note that lysine and arginine residues in the enzyme are involved in binding.)

5. Nitrites in food have been a concern in recent years because they lead to the formation of nitrosamines, which appear to be carcinogenic. Additionally, nitrites can be converted to nitrous acid by stomach HCl. Nitrous acid can oxidize amino groups to keto groups on the bases of DNA and RNA. Write reactions showing the effects of nitrous acid on cytosine and adenine in DNA. Would these changes in DNA damage the cell? Explain.

DNA Replication, Repair, and Recombination

I. KEY TERMS

With the help of this study guide, your textbook, and class notes, you should be able to define and explain the significance of the following terms:

direct repair
distributive
DNA polymerase
excision repair
general excision-repair pathway
genetic recombination
helicase

Holliday junction
homologous recombination
lagging strand
leading strand
nick translation
Okazaki fragment
photoreactivation
primase

primosome
processive
replication
replication fork
replisome
RNA primer
semiconservative

II. EXERCISES

A. True-False

_____ 1. The conservative model of DNA replication is supported by experimental evidence.

_____ 2. In *E. coli*, replication is unidirectional.

_____ 3. Bacterial replication requires dAMP, dTMP, dGMP, and dCMP.

_____ 4. Both eukaryotic and prokaryotic cells have several different DNA polymerases that are involved in replication and repair of DNA.

_____ 5. Kornberg's enzyme (DNA polymerase I) catalyzes the synthesis of DNA if provided with only the four deoxyribonucleoside triphosphates.

_____ 6. DNA polymerase III performs the major role in prokaryotic replication.

_____ 7. DNA polymerase III requires only a single-stranded DNA template for replication.

_____ 8. The reaction by which deoxyribonucleotides are added during replication is a near-equilibrium process.

_____ 9. Eukaryotic Okazaki fragments are smaller than prokaryotic Okazaki fragments.

_____ 10. DNA ligase plays a role in normal DNA replication as well as in repair of damaged DNA.

_____ 11. The correction of thymine dimer formation in DNA strands requires excision of the damaged region and its replacement by a new DNA segment of the same base sequence as that removed.

_____ **12.** Requirements of normal recombination in *E. coli* are that a region of double-stranded DNA be unwound to create a single-stranded section and a segment of duplex DNA be available with a sequence the same as or similar to that of the single-stranded section.

_____ **13.** Because eukaryotic genomes are so much larger than those of prokaryotes, replication times are comparatively much longer than those of prokaryotes.

B. Short Answers

1. The location at which duplex DNA is unwound during replication and at which new DNA is being synthesized is called the _____.

2. During replication, proteins that bind to individual unwound DNA strands to prevent them from reforming as a double helix are called _____.

3. The DNA strand that serves as a complementary pattern along which a new DNA strand is synthesized is called the _____ strand.

4. Short segments of RNA required for synthesis of DNA on a template are called _____.

5. The DNA strand formed as a continuous strand during replication is called the _____ and the strand formed in a discontinuous process is called the _____.

6. Enzymes that remain bound to their polymeric chains (e.g., DNA) during the process of many polymerization steps are said to be _____.

7. Short segments of lagging-strand DNA formed by the discontinuous process are called _____.

8. _____ is the name of the process by which DNA polymerase I catalyzes the formation of DNA segments to replace the RNA primer segments required by DNA polymerase III during replication.

9. In the process of general recombination, the structure formed as a result of exchange of single strands (strand invasion) is called the _____.

10. The protein complex in *E. coli* responsible for DNA replication is called the _____.

11. Since DNA polymerase III requires single-stranded DNA as a template, duplex DNA must be unwound by proteins called _____.

12. The enzyme that catalyzes the formation of RNA primers during replication is called _____ and is part of a complex called the _____.

13. When DNA ligase of *E. coli* catalyzes the sealing of a nick in DNA, _____ acts as a cosubstrate.

14. Termination of replication in *E. coli* involves binding of a protein called _____ to the parent DNA. It prevents the replication fork from passing through the termination site by inhibiting the helicase activity of the replisome.

C. Problems

1. Bacterial circular DNA is said to have a contour length (end-to-end length of the stretched out native DNA molecule) of 1600 μm, but the length of the cell itself is only about 2 μm. How is this possible?

2. How does acid cause precipitation of DNA?

3. What is the direction of synthesis of new DNA strands?

4. What is the proofreading function of DNA polymerase III and how is it manifested?

5. How does the bacterial replisome synthesize both strands of DNA simultaneously if the strands are synthesized in a 5'→3' direction and the template strands of the parent DNA are antiparallel?

6. If RNA primers are required for DNA synthesis, why do DNA samples isolated after replication contain no RNA?

7. What are 2',3'-dideoxyribonucleoside triphosphates and for what are they used?

8. What are the advantages of the fact that DNA replication is processive?

9. Describe the general steps, similar in all organisms, by which damaged DNA is corrected by the excision-repair pathway.

10. Your text indicates that the synthesis of histones and of DNA occurs in different parts of the cell, but that they occur simultaneously, and that the amount of histone is equivalent to the amount of new DNA. Explain why this is necessary for proper cell function.

11. Given the overall error rate for polymerization and the chromosome size of *E. coli* (4.6×10^6 nucleotide pairs), what is the average number of incorrect bases expected to be incorporated in a single replication of the *E. coli* chromosome?

12. (a) What is different about the mechanism of action of *E. coli* DNA ligase from that of the DNA ligase in eukaryotic cells? (b) What kinetic mechanism does the enzyme employ? (c) What is the specificity of the DNA ligases with respect to DNA base sequences recognized?

D. Additional Problems

1. Indicate how prokaryotic Okazaki fragments could be distinguished from eukaryotic Okazaki fragments.

2. From the size of the *E. coli* chromosome (see Problem 11 of the previous section) and the size of the Okazaki fragments given in text Section 20.3.A., (a) calculate the number of Okazaki fragments (OF) formed during replication of one-third of the *E. coli* chromosome, and (b) calculate the time in minutes required for formation of the Okazaki fragments.

3. *E. coli* cells were provided with a short pulse of ^{3}H-deoxythymidine. Shortly thereafter, the cells were lysed, and the DNA was isolated and denatured. Gel electrophoresis of the denatured DNA indicated the presence of both long and short ^{3}H-labeled DNA fragments. (a) Is the ^{3}H-labeled DNA newly synthesized DNA or is it DNA that already existed in the cell? (b) Explain why the ^{3}H-labeled DNA fragments were not all of the same size.

4. Figure 20.1 shows the gel electrophoretic pattern for a short piece of DNA that is to be sequenced. The Sanger method, which employs the 2',3'-dideoxyribonucleoside triphosphates (ddNTPs), was used to obtain the gel. (See text Box 20.1.) The specific ddNTP used in each of the four trials is identified above each lane of the gel. The short primer used was 5'-TA-3'. Determine the sequence of both DNA strands of this DNA fragment being sequenced.

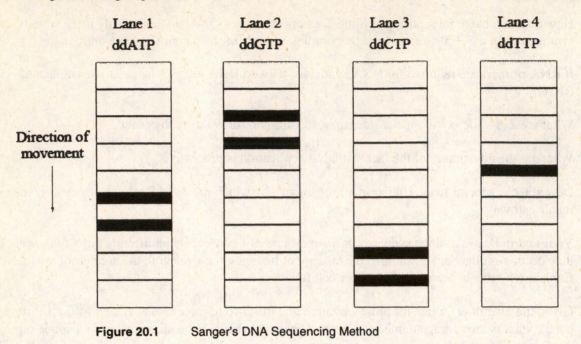

Figure 20.1   Sanger's DNA Sequencing Method

5. Using Figure 20.1 again, give the sequence, including the primer, of each fragment in each lane of the gel.

Transcription and RNA Processing

I. KEY TERMS

With the help of this study guide, your textbook, and class notes, you should be able to define and explain the significance of the following terms:

-35 region	intron	sense strand
activator	messenger RNA	sigma factor
antisense strand	open complex	small nuclear RNA
cap	operator	splice site
catabolite repression	operon	spliceosome
coding strand	pause site	splicing
consensus sequence	poly A tail	TATA box
constitutive expression	promoter	template strand
corepressor	repressor	termination sequence
exon	ribosomal RNA	transcription
gene	RNA polymerase holoenzyme	transcription bubble
inducer	RNA processing	transfer RNA

II. EXERCISES

A. True-False

_____ 1.   The direction of RNA synthesis is the same as that of DNA synthesis.

_____ 2.   The unusual bases, such as N^6-methyladenylate, inosinate, and dihydrouridylate, found in tRNA molecules are incorporated during transcription of the gene.

_____ 3.   Promoter regions are found exclusively on the DNA template strand.

_____ 4.   The RNA product of transcription is sequentially the same as that of the DNA coding strand, except that U replaces T.

_____ 5.   Not all genes that encode enzymes are inducible.

_____ 6.   Protein synthesis begins on prokaryotic mRNA molecules before synthesis of an mRNA molecule is complete.

_____ 7.   It is common for eukaryotic mRNA molecules to have poly A tails.

_____ 8.   Eukaryotic mRNA molecules are capped by the addition of a 7-methylguanylate group to the 3' end of the mRNA molecule.

_____ 9.   The 18S, 5.8S, and 28S rRNA molecules that are part of human ribosomes are all present in a single precursor rRNA molecule.

_____ 10. The σ^{70} subunit is a component of the RNA polymerase holoenzyme that recognizes some promoter sequences.

_____ 11. All transcription in typical eukaryotic cells is catalyzed by different RNA polymerases, found exclusively in the cell nucleus.

_____ 12. Some mature mRNA molecules are produced by splicing, which involves removal of noncoding segments from the transcript mRNA and joining of the remaining coding regions.

_____ 13. At a given moment, only about 3% of *E. coli* RNA is messenger RNA. This means that only a small amount of transcription activity is utilized to make mRNA.

_____ 14. None of the classes of RNA have catalytic activity.

_____ 15. CRP-cAMP is an activator complex specific for the *E. coli* lac operon.

B. Short Answers

1. The four classes of RNA include _____, _____, _____, and _____.

2. The most abundant class of RNA in the cell is _____.

3. The DNA region that acts as a transcription initiation signal is called the _____.

4. The prokaryotic holoenzyme that is involved in transcription is _____.

5. During transcription, double-stranded DNA is locally unwound. The resulting structure is called a/an _____.

6. _____ is the hexameric protein that aids in certain prokaryotic termination processes.

7. The eukaryotic enzyme that catalyzes the synthesis of mRNA precursors is _____.

8. During splicing of eukaryotic mRNA molecules, internal sequences called _____ are removed from primary RNA transcript. Sequences that are present in the primary RNA transcript and in the mature RNA molecule are called _____.

9. The RNA molecules that combine with proteins to form the spliceosome are called _____.

10. Transcription of a negatively regulated gene is prevented by a regulatory protein called a/an _____.

11. Transcription of a positively regulated gene requires the presence of a regulatory protein called a/an _____.

12. Ligands that bind to and inactivate repressors are called _____.

13. Genes that encode proteins or RNA molecules that are essential for the normal activities of cells are called _____.

C. Problems

1. How does the rate of RNA synthesis in *E. coli* compare with that of DNA?

2. What is the significance of the TATA box (-10 region) in bacterial DNA?

3. How does the regulatory protein rho help terminate transcription?

4. What is the *E. coli lacI* gene and what is its role in lactose use in *E. coli*?

5. What is catabolite repression?

6. What is CRP and what is the function of the CRP-cAMP complex?

7. What types of RNA are made by each of the eukaryotic RNA polymerases?

8. What is the function of capping of eukaryotic mRNA molecules?

9. The *E. coli* genome contains 4.6×10^6 base pairs (bp) and contains about 3000 genes that average 1500 bp each. This suggests that all but 2.2% of the bacterial genome is transcribed. How long would it take a single RNA polymerase to transcribe the expressed portion of the genome? (Assume that each gene is transcribed once and that initiation is not rate-limiting.)

10. (a) Why is the error rate of RNA synthesis higher than that for DNA replication? (b) Are transcription errors as serious to the cell as replication errors? Why or why not?

11. What are regions of dyad symmetry in DNA and how do they play a role in transcription?

12. Most bacterial mRNA molecules have half-lives of 2–3 minutes. Of what consequence is this with respect to the rate of transmission and the number of times mRNA is translated?

D. Additional Problems

1. Figure 21.1 shows data obtained from *E. coli* grown on a glycerol medium. The amount of mRNA molecules encoding lactose-metabolizing enzymes is plotted against time following the addition of isopropylthiogalactoside (IPTG) to the medium. Note that glucose is added approximately three minutes after IPTG addition. What two phenomena are illustrated by these data?

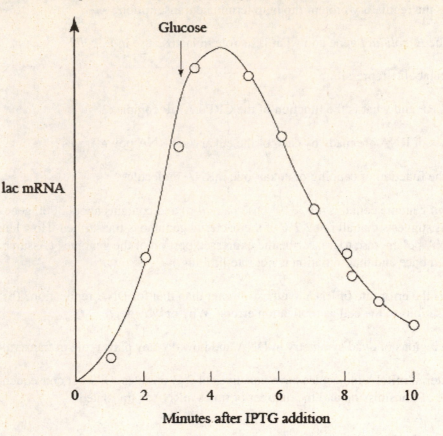

Figure 21.1 Effect of IPTG and Glucose on lac Operon

2. It has been estimated that about 10 repressor molecules (R) per *E. coli* cell are sufficient to prevent transcription of the *lac* operon. The dissociation constant for the repressor-operator complex (RO) has been estimated to be about 10^{-13} M.

$$RO \leftrightarrow R + O \qquad K_D = 10^{-13} \text{ M}$$

If the average cell has a volume of 0.30×10^{-12} mL and contains two copies of the *lac* operon, what is the concentration of operon (O) not bound by repressor at equilibrium?

3. Figure 21.2 shows a plot of data collected from T2-phage-infected *E. coli* in which the isolated RNA contains ^{32}P and the isolated DNA contains ^{3}H, allowing them to be separately detected. In this experiment, the DNA was denatured before being mixed with the RNA, and the components of the mixture were then separated by sedimentation in a CsCl density gradient. Note that the lower fraction numbers represent regions of higher density, removed first from the centrifugation tube. Hall and Spiegelman [*PNAS 47*, 141 (1961)] used this type of data to support the hypothesis that DNA serves as a template for RNA synthesis. Explain how these data support the hypothesis.

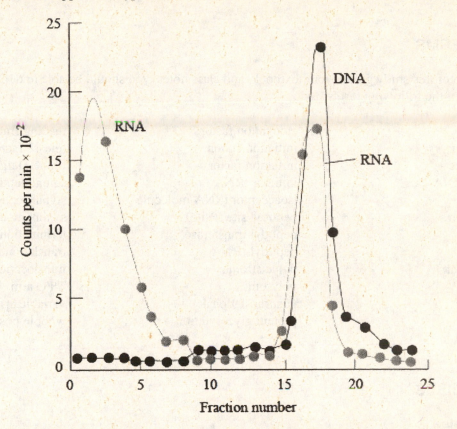

Figure 21.2 Separation of RNA and Denatured DNA by CsCl Density-Gradient Centrifugation

4. The *E. coli* core polymerase binds DNA nonspecifically with an association constant of 10^{10} M^{-1}. Binding to the promoter region is specific when the core polymerase combines with σ^{70} to form the holoenzyme. The association constant is then 2×10^{11} M^{-1}. With this tighter binding, how is the polymerase able to leave the promoter site and transcribe the rest of the DNA template?

5. Suppose occasional errors in processing of eukaryotic mRNA transcripts occurred in which an intron was not removed. What would be the likely result at the translational level?

Protein Synthesis

I. KEY TERMS

With the help of this study guide, your textbook, and class notes, you should be able to define and explain the significance of the following terms:

(A site)	elongation factor	reading frame
acceptor stem	initiation codon	release factor
aminoacyl site	initiation factor	Shine-Delgarno sequence
aminoacyl-tRNA	initiator tRNA	signal peptide
anticodon	isoacceptor tRNA molecule	signal-recognition particle
anticodon arm	peptidyl site (P site)	synonymous codon
attenuation	peptidyl transferase	termination codon
codon	peptidyl-tRNA	translational regulation
cotranslational	polycistronic	translocation
D arm	polysome	TΨC arm
degenerate	posttranslational	variable arm
E site	protein glycosylation	wobble position

II. EXERCISES

A. True-False

_____ 1. Every cell must contain at least 20 different tRNA species.

_____ 2. tRNA molecules are relatively small molecules composed of less than 100 nucleotides.

_____ 3. A particular aminoacyl-tRNA synthetase will recognize a specific amino acid, but will bind and aminoacylate any tRNA.

_____ 4. Text Equation 22.1 shows that ATP is required in the formation of aminoacyl tRNA molecules. This requirement simply provides energy for the process from the hydrolysis of a phosphoanhydride bond.

_____ 5. Protein synthesis always begins at the amino terminus in prokaryotic systems.

_____ 6. Transfer RNA molecules are charged by the covalent attachment of the amino acid for which they are specific at the 5' end of the tRNA.

_____ 7. N-Formylmethionyl-tRNA$_f^{Met}$ initiates protein synthesis in both prokaryotic and eukaryotic systems.

_____ 8. Base pairing between an aminoacyl-tRNA anticodon and an mRNA codon occurs in an antiparallel fashion.

_____ 9. In bacterial systems, the 30S ribosomal subunit binds somewhere other than the actual 5' end of the mRNA molecule to initiate protein synthesis.

_____ 10. Codons are translated in a 5'→3' direction.

_____ 11. Peptidyl transferase is a catalytic protein, found in the cell cytosol, which catalyzes peptide bond formation during translation.

_____ 12. Each new amino acid incorporated into a growing peptide chain furnishes the amino group used in the formation of a peptide bond.

_____ 13. Eukaryotic and prokaryotic mRNA molecules are polycistronic.

_____ 14. Prokaryotic mRNA molecules have Shine-Delgarno sequences for binding the small ribosomal subunit of ribosomes, but eukaryotic mRNA molecules do not.

_____ 15. Posttranslational modification of eukaryotic proteins to be secreted from the cell occurs after the proteins are secreted.

B. Short Answers

1. The only two amino acids with single codons are _____ and _____.

2. _____ is the term used to describe a family of tRNA molecules, each of which can be charged with the same amino acid.

3. The first codon to be determined, which coded for the amino acid _____, was _____.

4. When an aminoacyl tRNA synthetase acts to remove an amino acid mistakenly joined to a tRNA, the enzyme is exercising its _____ activity.

5. The double-stranded portion of the tRNA molecule where an amino acid is attached as the tRNA is charged is called the _____.

6. In some cases, several codons, called _____, encode the same amino acid.

7. The process by which the ribosome is shifted by one codon relative to mRNA is called _____.

8. A collection of ribosomes bound to a common mRNA template is called a/an _____.

9. The activation and incorporation of an amino acid into a growing peptide chain requires energy equivalent to the hydrolysis of _____ phosphoanhydride bonds.

10. Proteins destined for secretion are synthesized with an N-terminal sequence, called the _____, which assists in transport of the protein into the lumen of the endoplasmic reticulum, as the protein is being synthesized. This sequence is later removed.

11. Prokaryotic proteins that recognize termination codons on an mRNA molecule and cause hydrolysis of the peptidyl-tRNA are called _____.

12. Modification of protein chains are categorized as either _____ or _____ with regard to when they occur.

C. Problems

1. How is an amino acid activated (made ready for incorporation into a protein)?

2. What is the difference between Class I and Class II aminoacyl-tRNA synthetases?

3. Your text indicates that about 1 in every 100 times, isoleucyl-tRNA synthetase catalyzes formation of valyl-adenylate rather than isoleucyl-adenylate because of the similarity of the side chains of the two amino acids. Is the error of incorporation of valine rather than isoleucine in a protein 1 in each 100 trials (i.e., 1%)?

4. When Nirenberg and Matthaei were elucidating the genetic code, they used the supernatant fraction from a homogenized bacterial system that contained only ribosomes, tRNAs, and cytoplasmic enzymes. They added ATP, GTP, radioactive amino acids, and viral RNA. The system produced radioactive proteins except when the viral RNA was omitted. Explain.

5. Assume you had performed experiments to elucidate the genetic code. You made a synthetic RNA molecule that contained only adenine bases except for the terminal 3' base, which was cytosine. Using this synthetic mRNA molecule along with tRNAs, amino acids, and cytoplasmic enzymes, you found that the protein produced contained all lysine residues with the exception of the C-terminal residue, which was asparagine. Assuming a triplet code, but with no code words known at that time, state three or four conclusions you could reach from the results of your experiment.

6. Is there an advantage to bacterial systems that have, during transcription, mRNA synthesized from the 5'-end toward the 3'-end and that undergo translation in a 5'→3' direction along a template mRNA molecule?

7. What does it mean to say that the genetic code is degenerate?

8. Suppose you had been involved in determining the way in which the correct amino acids were incorporated into proteins. You wished to determine if it was the amino acid itself that was recognized by the codon of the mRNA or if it was the tRNA carrier that was recognized. To do this, you prepared cysteinyl-tRNA using ^{14}C-cysteine and cysteinyl-tRNA synthetase. You chemically transformed the radioactive cysteinyl group into an alanyl group and used the modified aminoacyl-tRNA in the synthesis of hemoglobin. After synthesis, the hemoglobin was examined for the location of radioactive residues. (a) What results would have occurred if the amino acid residue was recognized by the codon? (b) What results would have occurred if the carrier tRNA was recognized by the codon?

9. N-Formylmethionyl-tRNA$_f^{Met}$ initiates protein synthesis in many bacterial cells. (a) What prevents f-Met from being incorporated in protein chains at AUG codons at locations other than at the initiation site? (b) What prevents methionyl-tRNA^{Met} from initiating protein synthesis?

10. In prokaryotes, after the first peptide bond is formed between N-formylmethionine and the next amino acid that is brought into the A site, (a) how is the protein synthesis machinery made ready for the next (third) amino acid to be incorporated? (b) What is the name of this process?

11. Contrast translation in eukaryotic systems with that in prokaryotic systems.

12. Calculate the energy in kJ required for the synthesis of 0.685 mol of a protein that contains 250 amino acid residues, by the translation process, assuming standard conditions.

13. List several specific types of posttranslational processing of peptide chains that are known to occur, although not necessarily to each and every protein synthesized.

D. Additional Problems

1. In a particular bacterial cell, it is estimated that a single translation complex can synthesize a 300-residue polypeptide in about 20 seconds. The cell is estimated to contain about 5000 mRNA molecules at a given moment and is capable of producing about 1000 protein molecules per second. (a) For synthesis of the 300-residue protein by a single translation complex, how many mRNA nucleotides are being translated per second? (b) Based on the estimated cell production capacity, how many translation complexes, on average, are translating a particular mRNA at any given moment?

2. You are investigating the effect of various compounds on protein synthesis. You find that oligonucleotides containing Shine-Delgarno sequences inhibit prokaryotic systems but not eukaryotic systems. (a) What particular translation process was being inhibited? (b) Why are both types of systems not inhibited? (c) Would truncated proteins be formed in the prokaryotic system?

3. Assume you had synthesized an mRNA molecule that contains only uracil and guanine residues. The nucleotides were randomly incorporated and analysis showed that 76% of the bases in the mRNA were uracil and 24% were guanine. (a) What amino acid residues, and what percentages of each, would be incorporated into a protein synthesized using this molecule as a messenger? (b) If the protein contained 1000 residues, how many of the most abundant residues would there be? (c) How many of the least abundant residues would there be?

4. Binding of signal-recognition particles (SRP) to a segment of some incompletely formed proteins prevents further translation. (a) From what is the name for the SRPs derived? (b) What happens to allow the completion of the partially synthesized protein?

5. Selenium is an essential trace element found in both prokaryotic and eukaryotic enzymes. Selenium is present in certain *E. coli* enzymes as selenocysteine residues, formed by the selenation of serinyl-tRNA^{Ser}, catalyzed by serinyl-tRNA^{Ser} synthetase. This modified aminoacyl-tRNA recognizes the codon, UGA. (a) Draw a possible structure for the selenocysteine residue. You may represent the tRNA as in text Figure 22.9. (b) What is unusual about the codon recognized by the aminoacyl-tRNA?

Recombinant Technology

I. KEY TERMS

With the help of this study guide, your textbook, and class notes, you should be able to define and explain the significance of the following terms:

cDNA	marker gene	selection
chromosome walk	polymerase chain reaction	shuttle vector
cloning	probe	site-directed mutagenesis
cosmid	probing	sticky end
DNA library	recombinant DNA molecule	transfection
expression vector	recombinant DNA technology	transgenic organism
genetic transformation	restriction fragment length	vector
insert DNA	polymorphism	
insertional inactivation	screen	

II. EXERCISES

A. True-False

_____ 1.   Plasmids are circular, single-stranded, supercoiled molecules of DNA.

_____ 2.   Recombinant DNA molecules are inventions of man.

_____ 3.   A vector becomes a recombinant DNA molecule once insertion of a desired DNA fragment is accomplished.

_____ 4.   The normal flow of information goes from DNA to mRNA to protein, but it is possible to make DNA from the information provided by a specific mRNA molecule.

_____ 5.   Some vectors can replicate in either prokaryotic or eukaryotic cells.

_____ 6.   If a DNA vector and the DNA fragment to be inserted have complementary sticky ends, a stable recombinant DNA molecule can be created by simply annealing the two types of DNA.

_____ 7.   Some cells take up recombinant DNA molecules directly by a process called conjugation.

_____ 8.   Insertion of a DNA fragment into the amp^R gene of the pBR322 plasmid would lead to transformants that are resistant to both ampicillin and tetracycline.

_____ 9.   In the preparation of phage particles containing recombinant DNA molecules, the λ phage DNA is cleaved with a restriction endonuclease to yield λ arms and stuffer fragment(s). The stuffer fragment(s) is(are) purified so it(they) can be combined with genomic DNA to create the recombinant molecule.

_____ **10.** The poly A tails of mature eukaryotic mRNA molecules are important to the formation of cDNA molecules using reverse transcriptase.

B. Short Answers

1. A synthetic DNA fragment made to include a recognition sequence to allow insertion of a target DNA molecule into a vector is called a/an _____.

2. The direct uptake of foreign DNA by a host cell is called _____.

3. _____ is the process that uses a pulse of high voltage to increase cell permeability to allow uptake of recombinant DNA by cells.

4. A collection of bacterial colonies, each containing a different duplex DNA segment inserted into a plasmid is called a/an _____.

5. _____ involves introduction of a normal gene into mutant cells of an individual in an effort to correct a genetic disease.

6. A _____ is a molecule that specifically recognizes a desired DNA segment in a large library.

7. Plasmids that have been modified to contain eukaryotic promoters, ribosome-binding sites, and transcription terminators are called _____, since they allow the expression of eukaryotic genes of recombinant DNA molecules in prokaryotic cells.

8. An animal or plant that has been engineered to contain stably-integrated foreign DNA is said to be _____.

9. Variations that occur in the pattern of restriction fragments, produced by the same restriction endonuclease, from the DNA of different individuals of the same species are known as _____.

10. Amplification of small amounts of DNA taken from a source such as a hair follicle is accomplished by a technique called the _____.

11. _____ is the technique by which a gene of a vector is rendered incapable of expression of the gene product due to the insertion of a foreign DNA fragment into the gene.

12. The introduction of recombinant DNA into a host cell by engineered phage particles occurs by a process called _____.

13. The amp^R gene, used in some vectors to allow identification of host cells that contain the recombinant DNA molecules, is called a/an _____.

14. The process by which a particular nucleotide in a gene is replaced by another is called _____.

C. Problems

1. List the six steps outlined in your text common to most recombinant DNA experiments.

2. Name four different types of vectors that have been used in recombinant DNA technology.

3. Without using enzymes, how could a cDNA-mRNA hybrid be easily treated in the laboratory to yield the cDNA as a single strand?

4. What action, if any, would occur with an mRNA-cDNA hybrid in the presence of RNase H?

5. Indicate two advantages to the use of λ phage as a vector rather than a plasmid.

6. (a) Why must λ phage DNA be engineered to be a suitable vector for recombinant DNA expression? (b) What is done to the phage DNA prior to using it to make recombinant DNA molecules?

7. In the production of cDNA using reverse transcriptase, why is DNA polymerase I used in the synthesis of duplex DNA rather than DNA polymerase III?

8. In what way are cDNA libraries preferable to genomic libraries?

9. To create cDNA libraries, mRNA molecules must be isolated and purified. How can these molecules be separated from the many other molecules of rRNA and tRNA?

10. Suppose a researcher was interested in determining whether or not particular amino acid residues of an enzyme were involved in the catalytic activity characteristic of the enzyme. What genetic engineering approach is now used in such studies?

11. The polymerase chain reaction (PCR) is used to amplify small amounts of DNA. Duplex DNA is heated to separate its strands for synthesis, since single DNA strands are required to serve as templates. (a) Since elevated temperatures denature not only DNA, but also enzymes and other proteins, how is synthesis of the new strands of DNA accomplished? (b) Cite three or four circumstances in which the use of PCR allows investigators to obtain knowledge that would not be possible otherwise.

12. In some cases, mechanical shearing of DNA creates fragments with blunt ends. Usually, researchers want DNA fragments with sticky ends for construction of recombinant DNA molecules. Phage T4 DNA ligase is an enzyme that will catalyze the covalent joining of *any* two DNA molecules with blunt ends. How might one use this information to modify a DNA fragment that would contain a particular restriction site that would produce sticky ends when cleaved by the specific restriction endonuclease?

13. As previously mentioned, plasmid vectors have been engineered to allow expression of eukaryotic genes in prokaryotic cells. A hybrid promoter, p_{tac}, has been made that contains part of the *lac* promoter and part of the *trp* promoter. This hybrid promoter must be activated by isopropyl-β-thiogalactoside (IPTG) before it will drive expression of inserted foreign genes. If production of a eukaryotic protein is the desired end, why engineer a plasmid with a promoter that must be activated?

14. Consider the synthesis of cDNA. Explain how you would know that reverse transcriptase synthesizes the single DNA strand of the mRNA-cDNA hybrid in a 5'→3' direction.

15. Indicate an advantage of cDNA libraries compared with genomic libraries.

D. Additional Problems

1. Itakura and coworkers tried to produce the hormone somatostatin by recombinant DNA techniques. Initial attempts using the transformed bacterial cell were unsuccessful. The investigators then attached the somatostatin gene to the bacterial *lac* gene with a bridge between the two that consisted of the base sequence ATG in the template strand. Expression was successful. Why was the bridge sequence used? (Note that the somatostatin gene does not contain an ATG sequence.)

2. Radioactive probes, such as a DNA segment complementary to a single-strand region of interest, are used to locate particular genes. Such complementary DNA (cDNA) sequences can be synthesized if the corresponding mRNA sequence is known. The problem is complicated, however, if only the sequence of the protein for which the mRNA is the blueprint is known. Explain.

3. With reference to the previous problem, how many different DNA probes would be necessary to find the DNA template strand that corresponds to the mRNA segment that gives rise to the peptide segment Trp–Phe–Lys–Glu–Met?

4. Write sequences of the DNA probes from the previous problem.

5. It is estimated that a 1 kb fragment of human DNA represents about 0.000031% of the human genome. What, then, is the size of the human genome?

6. How many clones must be screened, with a probability of 0.99 that a 10 kb fragment was located, from a genome of 144,000 kb?

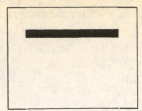

Solutions

Chapter 2

A. True-False

1. <u>False</u>.   The highly electronegative *oxygen* pulls the shared electrons closer to the oxygen than the hydrogen, giving rise to the polar bond. The electronegativity of hydrogen is considerably less than that of oxygen.

2. <u>True</u>.   Because the H–O–H bond angle is 104.5°, the two polar O–H bonds result in a center of negative charge for the molecule that does not coincide with the center of positive charge as indicated in text Figure 2.2a.

3. <u>True</u>.   Potential hydrogen bonds can involve the two hydrogen atoms of H_2O and the two nonbonding electron pairs on the oxygen atom. This possibility is illustrated in text Figures 2.4 and 2.5. Some sources indicate, however, that an average of 3.4 hydrogen bonds form in liquid water.

4. <u>True</u>.   Electrolytes (like NaCl) are polar compounds that dissociate into their anions and cations, each surrounded by water molecules that form a solvation sphere. Nonelectrolytes (like glucose and ethanol) also become hydrated by water molecules due to hydrogen bond formation between water and the polar – OH groups of the solvated molecules.

5. <u>False</u>.   The osmotic pressure depends on the total molar concentration of solute, not on the chemical nature of the solute(s).

6. <u>True</u>.   The dipoles thus produced are short-lived and the strengths of the induced interactions are low.

7. <u>True</u>.   At pH 3.0, however, the $[H^+] > [OH^-]$.

8. <u>False</u>.   An acid is weak if it dissociates less than 100% in water.

9. <u>True</u>.   Use the pKa for lactic acid from text Table 2.4 and employ the Henderson-Hasselbalch equation. When $[A^-]/[HA] = 1$, the log $[A^-]/[HA] = 0$, and pH = pK_a of the weak acid.

10. <u>True</u>.

11. <u>False</u>.   It is called a solvation sphere.

12. <u>True</u>.   Compare van der Waals forces of 0.4 kJ mol^{-1} in methane to the hydrogen bonds between the bases in DNA that are 2-7.5 kJ mol^{-1}.

13. <u>True</u>.   Greater $[A^-]/[HA]$ leads to larger $log[A^-]/[HA]$ values in the Henderson-Hasselbalch equation, thus, larger pH values.

B. Short Answers

1. Electronegativity

2. Chaotropic

3. Amphipathic

4. micelles

5. solvation sphere

6. van der Waals forces

7. M^2 or $(moles/liter)^2$

8. weak

9. M or moles/liter

10. buffer

11. acidosis

12. electrolytes

13. hemoglobin

C. Problems

1. The two nonbonding electron pairs strongly repel each other, thus pushing the electron pairs of the two O-H covalent bonds closer together than might be expected.

2. Molecular crowding is the most important factor that leads to alteration of diffusion of molecules within cells.

3. HCl, a strong acid, ionizes 100%.

$$HCl \rightarrow H^+ + Cl^-$$

Therefore, $[H^+] = [Cl^-] = 1.00 \times 10^{-2}$.

$$pH = -\log (1.00 \times 10^{-2}) = 2.00.$$

Acetic acid is weak. It does not ionize 100%.

	CH_3COOH	$\leftrightarrow$	H^+	+	CH_3COO^-
Initial:	0.0100 M		0		0
At Equilibrium:	0.0100 - x		x		x

where x represents the moles/liter of acetic acid that ionize and the moles/liter of H^+ and of CH_3COO^- present at equilibrium due to that ionization. You can neglect x in the term 0.0100 - x when the acid $K_a <$ about 10^{-4} and when the original concentration of the weak acid is > about 10^{-2} M. Making this assumption here, the equilibrium concentration of acetic acid is: $[CH_3COOH] = 0.0100 - x = 0.0100$.

$$K_a = \frac{[H^+][CH_3COO^-]}{[CH_3COOH]}$$

From text Table 2.4, the K_a for acetic acid is $= 1.76 \times 10^{-5}$. Therefore,

$$1.76 \times 10^{-5} = \frac{(x)(x)}{0.0100}$$

$$x^2 = (0.0100)(1.76 \times 10^{-5}) = 1.76 \times 10^{-7}$$
$$x = [H^+] = 4.20 \times 10^{-4}$$
$$pH = -\log (4.20 \times 10^{-4}) = 3.38.$$

The HCl puts more H^+ ions in solution since it is 100% ionized. Acetic acid is less than 100% ionized and releases fewer H^+ ions, which results in a higher pH.

4. HNO_3 is a strong acid, so $[H^+] = 3.50 \times 10^{-3}$.

$$pH = -\log 3.50 \times 10^{-3} = 2.46.$$
$$K_w = [H^+][OH^-] = 1.00 \times 10^{-14}; \text{ therefore,}$$
$$[OH^-] = (1.00 \times 10^{-14})/(3.50 \times 10^{-3}) = 2.86 \times 10^{-12}.$$

5. The K_a for acetic acid is 1.76×10^{-5}. The pK_a is 4.75. Note that, since the pH is lower than the pK_a, one should expect a greater acid than conjugate base concentration (i.e., [CH3COO-] < [CH3COOH]). Use the Henderson-Hasselbalch equation:

$$pH = pK_a + \log [CH_3COO^-]/[CH_3COOH]$$
$$\log [CH_3COO^-]/[CH_3COOH] = pH - pK_a$$
$$[CH_3COO^-]/[CH_3COOH] = \text{antilog } [pH - pK_a]$$
$$[CH_3COO^-]/[CH_3COOH] = \text{antilog } [4.25 - 4.75]$$
$$[CH_3COO^-]/[CH_3COOH] = 0.316. \text{ Thus, the ratio of acetate to acetic acid is 0.316 to 1 or 1 to 3.16.}$$

6. If 23 kJ represents the strength of each of the four hydrogen bonds present in ice, 78.2 kJ would indicate that fewer than four hydrogen bonds are present in liquid water. The average number of hydrogen bonds is:

$$\frac{78.2 \text{ kJ}}{\text{molecule}} \times \frac{1 \text{ hydrogen bond}}{23 \text{ kJ}} = \frac{3.4 \text{ hydrogen bonds}}{\text{molecule}}$$

7. H_3O^+ and H_2O are a conjugate acid-base pair, but so are H_2O and OH^-. Water is amphoteric, which means it can act both as an acid and as a base.

8. If you jump to your calculator immediately without thinking about this problem, you will probably answer incorrectly. The pH is not 8. That is a basic pH. In this dilute solution, the H^+ from the ionization of water is the major contributor to the pH. $[H^+]$ from water is 1.0×10^{-7} and that from the HCl is 10^{-8} or 0.1×10^{-7}. Use the sum of the two concentrations, 1.1×10^{-7}, to determine the pH.

$$pH = -\log 1.1 \times 10^{-7} = 6.958 = 7.0, \text{ to two significant figures. Not very acidic, is it?}$$

9. For weak acids at low concentrations,

$$[H^+] = (K_a [HA])^{1/2}$$
$$[H^+] = [(1.37 \times 10^{-4})(0.0105)]^{1/2} = 1.20 \times 10^{-3}.$$
$$pH = -\log 1.20 \times 10^{-3} = 2.92.$$

10. At this point in the titration, half of the lactic acid molecules have had their protons removed (titrated) by OH- ions of the added base, forming water and an equal amount of lactate ion. With one-half the lactic acid still present in protonated form, there are equal amounts of a weak acid (lactic) and its conjugate base (lactate ion). This is a buffer system. The pH is the same as the pK_a of the acid, since the Henderson-Hasselbalch equation term $\log [A^-]/[HA] = 0$. Therefore,

$$pH = pK_a + 0 = 3.86.$$

11. $pH = pK_a + \log([A^-]/[HA])$
    $= 3.19 + \log [(0.0400)/(0.0350)]$
    $= 3.19 + \log 1.14$
    $= 3.19 + .06 = 3.25$

12. H_2CO_3 forms rapidly by reaction of dissolved carbon dioxide with water.

$$H_2O + CO_2 \leftrightarrow H_2CO_3$$

As H_2CO_3 is used, more is formed from the carbon dioxide carried in the blood, thus assuring an adequate supply.

D. Additional Problems

1. pH = pK_a + log $[NH_3]/[NH_4^+]$
   = 9.2 + log $[(0.025)/(0.030)]$
   = 9.2 + log 0.83
   = 9.2 + (-0.08)
   = 9.12 or 9.1 to one decimal place.

2. If acetic acid were a strong acid, the $[H^+]$ would be 1.00×10^{-2}. However, $[H^+]$ was found to be 4.20×10^{-4}. Thus, % ionization = $(4.20 \times 10^{-4}/1.00 \times 10^{-2}) \times 100 = 4.20\%$.

3. The pK_{a1} for H_2CO_3 is 6.37. Use this in the Henderson-Hasselbalch equation.

$$pH = 6.37 + \log ([HCO_3^-]/[H_2CO_3]); \text{ and,}$$
$$[HCO_3^-]/[H_2CO_3] = \text{antilog} (7.28 - 6.37). \text{ Therefore,}$$
$$[HCO_3^-]/[H_2CO_3] = 8/1.$$

4. If $K_w = 1.4 \times 10^{-14}$ M^2, $[H^+] = (1.4 \times 10^{-14})^{1/2} = 1.18 \times 10^{-7}$. Then, pH = $-\log (1.18 \times 10^{-7}) = 6.93$. The solution is neutral because it contains equal quantities of H^+ and OH^-, since one of each is formed when an H_2O molecule ionizes.

$$H_2O \leftrightarrow H^+ + OH^-$$

This illustrates that the familiar pH scale is based on a specific temperature (25 °C).

5. In the Henderson-Hasselbalch equation, the quantities of A^- and HA are expressed as moles/liter. Since both species are in the same solution and thus have the same volume, the volumes in the $[A^-]/[HA]$ ratio cancel. A ratio of the moles of A^-/HA remains.

Moles A^- (HPO_4^{2-}) = 0.4000 L $\times$ 0.12 mol/L = 0.048.
Moles HA ($H_2PO_4^-$) = 0.6000 L $\times$ 0.075 mol/L = 0.045.
pH = 7.21 + log $[(0.048)/(0.045)]$
     = 7.21 + 0.028 = 7.238 = 7.24.

6. Assume that all dissolved CO_2 forms H_2CO_3:

$H_2O + CO_2 \rightarrow H_2CO_3$.
If pH = 5.68, $[H^+] = 2.09 \times 10^{-6}$. Also,
$[H^+] = ([HA] K_a)^{1/2}$ or $[H^+]^2 = [HA] K_a$. Then,
$[HA] = [H^+]^2/K_a = (2.09 \times 10^{-6})^2/(4.3 \times 10^{-7})$
        = 1.02×10^{-5} M.

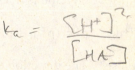

$k_a = \dfrac{[H^+]^2}{[HA]}$

7. From the Henderson-Hasselbalch equation,

$$7.30 = 7.2 + \log [(x\ K_2HPO_4/L)/(0.0800\ mol\ KH_2PO_4/L)]$$
$$[(x\ K_2HPO_4/L)/(0.0800\ mol\ KH_2PO_4/L)] = antilog\ (7.30 - 7.20)$$

x K_2HPO_4 =(0.0800)(1.26) = 0.101 mol. Therefore, since the molar mass of K_2HPO_4 is 174.18 g/mol,

g K_2HPO_4 = 0.101 mol × 174.18 g/mol
            = 17.6 g.

Chapter 3

A. True-False

1. <u>True</u>.   The α-amino group (pK_a = 9.9) of all L-aspartic acid molecules will be protonated, $-NH_3^+$, as will half the α-carboxyl groups, –COOH. The other half will be ionized ($-COO^-$), since the pK_a for that group is 2.0. Since the pK_a for the side chain carboxyl group is 3.9, few of those groups will be ionized and the majority of the molecules will then carry a net positive charge.

2. <u>True</u>.   Use the Henderson-Hasselbalch equation to determine the amount of conjugate base to acid form for each of the three ionizable groups at pH 10. On the basis of 100 lysine molecules:

    α-amine group: 11 NH_3^+, 89 NH_2
    α-carboxyl group: 100 COO^-, 0 COOH
    amine of R-group: 76 NH_3^+, 24 NH_2

The probability of a form with three separate charges is:

$$p = 11/100 \times 100/100 \times 76/100 = 0.0836 = 0.084$$

So, about 8.4% of the molecules have three separate charges—not the predominant form. (You may want to try this at other pH values.)

3. <u>False</u>.   Only 7 of the 20 amino acids are classified as highly hydrophilic.

4. <u>True</u>.

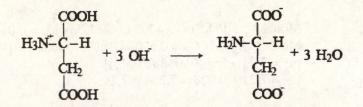

5. <u>False</u>.   Glycine is not. It has no chiral center since its side chain is an H atom.

6. <u>True</u>.   pI = (pK_{a1} + pK_{a2})/2. (Try calculating pI for L-alanine, L-leucine, and L-valine.)

7. <u>False</u>.   The direction of rotation of polarized light must be determined empirically. L- refers to the structural similarity to L-glyceraldehyde.

8. <u>False</u>.   The solution would show no optical activity because the equal and opposite optical activities of the D- and L-valine components would cancel. Equal amounts of two enantiomers constitute a racemic mixture.

9. <u>False</u>.   It is a pentapeptide. It is the number of amino acid residues that dictates the prefix used. Five residues would involve four peptide bonds, as follows: A–B–C–D–E (the dashes (–) represent peptide bonds).

10. <u>False</u>.   Cysteine is the only exception. L-cysteine is R-cysteine because the side chain, $-CH_2SH$, has a higher priority than the carboxyl group, $-COOH$. See Figure 3.1.

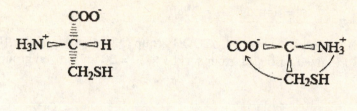

R—designation

Figure 3.1   The RS Designation of L-Cysteine

11. <u>False</u>.   A pH of 5.7 is the isoelectric point for methionine, so methionine will not migrate. Aspartic acid will have a net negative charge at pH 5.7 and will migrate toward the anode.

12. <u>True</u>.   SDS denatures proteins and associates with protein molecules giving all molecules a negative charge proportional to their length. Thus, the charge/mass ratios are the same for all SDS-proteins. Migration rates depend on size and charge/mass ratios. Since proteins coated with SDS anions have the same charge/mass ratio, they will migrate according to size only.

13. <u>False</u>.   Acid hydrolysis converts L-glutamine and L-asparagine to L-glutamic and L-aspartic acid, respectively. L-Tryptophan is destroyed in such acidic conditions and some L-serine, L-threonine, and L-tyrosine are also lost.

14. <u>True</u>.   Cloning and sequencing of DNA is often easier, due to new technology, than is the purification and sequencing of the protein itself.

B. Short Answers

1. eluted

2. zwitterion

3. isoelectric point

4. disulfide bridge

5. hydropathy

6. microenvironment

7. enantiomers

8. primary structure

9. Edman degradation procedure; sequenator

10. N-terminus; C-terminus

11. Arginine

12. charge; mass

C. Problems

1. The pK_a for the α-carboxyl group of alanine is 2.4. Use the Henderson-Hasselbalch equation to determine the ratio of conjugate base groups (–COO⁻) to acid groups (–COOH).

$$pH = pK_a + \log [A^-]/[HA]; \text{ let } X = [A^-]/[HA].$$
$$\log X = pH - pK_a = 3.0 - 2.4 = 0.6$$
$$X = \text{antilog } 0.6 = 3.98.$$

So, there are 4 –COO⁻ groups for every 1 –COOH group (5 total groups). Per 100 alanine molecules, there are $4 \times 20 = 80$ ionized carboxyl groups and $1 \times 20 = 20$ un-ionized groups.

2.

Note that, although these forms are enantiomers of *each other*, each is a diastereomer of the form given in the question.

3. Although glycine has a relatively low molecular weight (75 g/mol), it exists in a salt form (zwitterionic form) and has properties similar to those of other salts, namely stability of crystals to relatively high temperature. Amino acids typically decompose over a relatively high temperature range.

4.

5. See text Table 3.2 for pK_a values of serine.

Note: These are the predominant forms in each population.

6. (a) At pH 3.0, virtually all of the N-terminal amino groups will be protonated (–NH₃⁺). By use of the Henderson-Hasselbalch equation, you can determine that only 11 of every 100 molecules of aspartame will have the carboxyl group of the aspartate side chain ionized (–COO⁻). No other groups are capable of ionizing since the C-terminus exists as a methyl ester. Therefore, the predominate ionic form of aspartame at pH 3.0 is positively charged and will migrate toward the negative electrode, the cathode.

(b) Use the pK_a values in text Table 3.2 for the α-amino group and the side chain carboxyl group of L-aspartate to calculate the isoelectric point of aspartame. It is 6.9, so the aspartame sample should not migrate in an electric field at pH 6.9. Note again that no other groups can contribute to the ionic form of the molecules at this pH.

7.

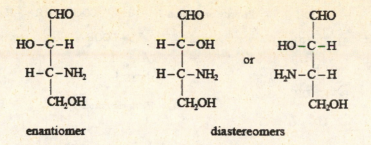

enantiomer   diastereomers

8. You will find pK_a values for these and the other amino acids in Table 3.2 of your text.

L-Valine: pI = (2.3 + 9.7)/2 = 6.0

L-Lysine: pI = (9.1 + 10.5)/2 = 9.8

L-Aspartic Acid: pI = (2.0 + 3.9)/2 = 3.0

Note that, for L-lysine, only the two like (basic) groups were averaged because the pK_a values of these groups lie equidistant on either side of the pH at which there exists a species with a net zero charge. The two like groups of L-aspartate are the two acidic groups.

9. Figure 3.2 shows the titration curve of an amino acid with a pK_{a1} of 2.4 and a pK_{a2} of 9.6. Note that one equivalent of OH^- titrates one equivalent of –COOH converting it all to the –COO⁻ form. All the amino groups are still protonated (–NH₃⁺). This corresponds to the isoelectric point. The addition of 0.5 equivalent of OH^- titrates one-half of the total –COOH groups, resulting in equal numbers of –COOH and –COO⁻ groups. From the Henderson-Hasselbalch equation, the pH at this point is equal to the pK_a of the –COOH group. Similarly, when 1.5 equivalents of OH^- have been added, half of the protonated amino groups have been titrated and there are thus equal amounts of –NH₂ and –NH₃⁺ forms. The pH at that point is the pK_a of the amino group.

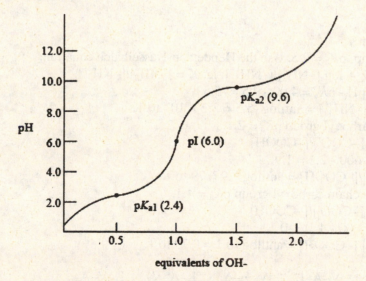

Figure 3.2   Titration Curve for an Amino Acid

10.

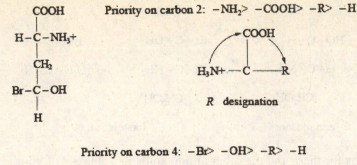

Priority on carbon 2: $-NH_2 > -COOH > -R > -H$

R designation

Priority on carbon 4: $-Br > -OH > -R > -H$

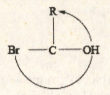

S designation

(2*R*, 4*S*)-2-amino-3-bromo-3-hydroxybutanoic acid

11. Cellular proteins are synthesized in an N-terminal to C-terminal direction, which is the direction by which the sequence of a protein is indicated.

12. Due to the association of SDS molecules, all proteins are negatively charged and travel toward the anode. Larger proteins encounter more resistance and migrate more slowly. Low molecular weight proteins travel faster toward the bottom of the gel.

D. Additional Problems

1. For the α-amino group, $pK_a = 9.5$. Use the Henderson-Hasselbalch equation:

$pH = pK_a + \log [-NH_2]/[-NH_3^+]$; let $X = [-NH_2]/[-NH_3^+]$.

$\log X = pH - pK_a = 4.00 - 9.5 = -5.5$

$[-NH_2]/[-NH_3^+] = $ antilog $-5.5 = 3.2 \times 10^{-6}$ to 1.

The α-carboxyl group pK_a 2.1.

Let $X = [-COO^-]/[-COOH]$

$\log X = 4.00 - 2.1 = 1.9$.

$[-COO^-]/[-COOH] = $ antilog $1.9 = 79$ to 1.

The side chain carboxyl group $pK_a = 4.1$.

Let $X = [-COO^-]/[-COOH]$

$\log X = 4.00 - 4.1 = -0.1$.

$[-COO^-]/[-COOH] = $ antilog $-0.1 = 0.79$ to 1.

2. A–V–L   A–L–V   V–A–L   V–L–A   L–A–V   L–V–A

3. There are 20 possibilities for each of 15 residue sites, or 20^{15}.

$$\log \text{answer} = \log 20^{15} = 15(\log 20) = 19.52;$$
$$\text{answer} = \text{antilog } 19.52 = 3.28 \times 10^{19}.$$

Note that this problem is different from the previous one since each peptide in that problem had to contain all three of the specified amino acids, A, V, and L.

4. Since the isoelectric pH of glycine is 6.1, it has very little tendency to move in the electric field. Having a slight excess of positive charge (since the pH is not quite at the isoelectric pH), it may travel slightly toward the negative electrode. The total time the electric field is applied is also a factor in extent of travel.

L-Lysine has a net positive charge since its $-NH_3^+$ groups tend not to release protons appreciably before pH 7–8. Lysine will migrate appreciably toward the negative electrode. L-Aspartate has a net negative charge since both carboxyl groups are fully ionized ($-COO^-$) at pH 6.0. It will migrate toward the positive electrode.

5. The dipeptide composed of L-alanine and L-aspartic acid could have either of these primary structures:

H_3N^+–Ala–Asp–COOH   or H_3N^+–Asp–Ala–COOH

|

COOH   COOH

Each dipeptide has three groups that can donate protons. Thus, one mmole of dipeptide requires three mmole of base.

(30. mL of 0.1 M NaOH = 3 mmole NaOH).

The dipeptide composed of alanine and glycine could have either of these primary structures:

H_3N^+–Ala–Gly–COOH or H_3N^+–Gly–Ala–COOH

Each dipeptide has only two groups to react with base and would thus consume only 2 mmole of NaOH. The dipeptide that was titrated must then be the one composed of alanine and aspartic acid. (The primary structure is unknown).

6. (a) Cleavage by trypsin would yield two dipeptides of sequence:

Gly–Met–Arg + Asp–Phe–Gly

(b) Cleavage by BrCN would yield a dipeptide and a tetrapeptide of sequence:

Gly–Met* + Arg–Asp–Phe–Gly

(c) Cleavage by *S. aureus* V8 would yield a tetrapeptide and a dipeptide of sequence:

Gly–Met–Arg–Asp + Phe–Gly

(d) Cleavage by chymotrypsin would yield glycine and a pentapeptide of sequence:

Gly–Met–Arg–Asp–Phe + Gly

Note that Met* is used to indicate homoserine lactone, the residue created by BrCN action on L-methionine.

7. Note that use of a sequenator would provide a faster and less tedious way of sequencing such a small peptide. This problem, however, is designed to help you understand how larger polypeptides might have to be segmented by various cleaving agents in order to deduce their primary structures. To solve this problem, it may be helpful to make blanks for each of the amino acid residues from the N-terminus to the C-terminus and fill in the identities of each residue as deduced from the data. Equimolar amounts of Arg, Glu, Gly, Lys, Met, and Phe indicate that this is a hexapeptide.

N-term [( )-( )-( )-( )-( )-( )] C-term

(a) Since trypsin cleaves on the carbonyl side of L-lysine and L-arginine, the fact that L-arginine is found alone and the rest of the peptide is left intact suggests that L-arginine is N-terminus and L-lysine is C-terminus. If L-lysine were in the interior of the peptide, another cleavage would have occurred. So,

N-term [Arg-( )-( )-( )-( )-Lys] C-term

(b) Cyanogen bromide cleaves on the carbonyl side of L-methionine residues. Since this resulted in two tripeptides, the L-methionine must be the third residue from the N-terminus. You should note that the L-methionine residue would have been converted to homoserine lactone as indicated in your text. Thus,

N-term [Arg-( )-Met-( )-( )-Lys] C-term

(c) The *S. aureus* V8 protease cleaves on the carbonyl side of L-aspartate and L-glutamate residues. This action left free L-lysine and a pentapeptide. The L-lysine had to be on the carbonyl side of L-glutamate for this result to have occurred. This confirms that L-lysine is the C-terminal residue and that L-glutamic acid is the second residue from the C-terminus. Therefore,

N-term [Arg-( )-Met-( )-Glu-Lys] C-term

(d) Chymotrypsin cleaves on the carbonyl side of aromatic residues like L-phenylalanine. Since there are only two positions left available (2 and 4) in the hexapeptide, consider the possible results if phenylalanine were in each position. In position 2, cleavage would still give a dipeptide and a tetrapeptide, but the UV absorbance (260 nm) would be observed in the dipeptide and not the tetrapeptide. Therefore, L-phenylalanine must be in position 4. The only residue left is glycine and the only position left is position 2. The total primary structure is:

Arg-Gly-Met-Phe-Glu-Lys

8. SDS-PAGE results in separation of proteins in individual bands. Isolation of the protein from a particular band cut from the SDS-PAGE gel allows it to be used, even in picomole quantities, in mass spectrometry for very accurate determination of its molecular weight.

Chapter 4

A. True-False

1. <u>False</u>. Hydrogen bonds stabilize areas of secondary structure.

2. <u>True</u>. The *trans* conformation is spatially more favorable than the *cis* conformation. Rare exceptions occur, usually at peptide bonds involving the amide nitrogen of proline.

3. <u>True</u>. It is the native conformation that the protein normally has in its biological environment.

4. <u>True</u>. Enzymes are globular proteins that fold such that their active three-dimensional shape (native shape) is achieved.

5. <u>True</u>. Myoglobin is half saturated at a pO_2 of 2.8 torr, but a pO_2 of 26 torr is required for half saturation of hemoglobin.

6. <u>False</u>. The fact that the P_{50} for fetal hemoglobin is lower (18 torr) than that for adult hemoglobin is an indication that fetal hemoglobin binds O_2 more readily than does adult hemoglobin, thus making for efficient O_2 transfer to the fetus.

7. <u>False</u>. The capacities to bind oxygen are significantly different (due primarily to the lack of quaternary structure in myoglobin), but the primary structures of the protein chains are very similar.

8. <u>False</u>. The pitch is the distance along the α-helix that constitutes one turn of the helix and involves about 3.6 amino acid residues.

9. <u>False</u>. An isolated heme in solution does not bind oxygen. Instead, the Fe^{2+} ion of heme is oxidized to Fe^{3+}. This does not occur in hemoglobin or myoglobin molecules due to the microenvironment of the heme group in these proteins.

10. <u>False</u>. They are called motifs. Domains are independently folded regions and may consist of combinations of motifs.

11. <u>True</u>. This allows interaction of these water-loving groups with the water molecules that surround the protein.

12. <u>False</u>.   Normally, no covalent bonds are broken. Activity is lost, however, due to loss of the native tertiary structure, which is maintained by noncovalent interactions.

13. <u>True</u>.   Salt bridges (electrostatic attractions between positive and negative ions), hydrogen bonding, hydrophobic forces, and other interactions help maintain the association of the monomers of an oligomeric protein.

14. <u>False</u>.   Low pH (high $[H^+]$) causes a lowered oxygen affinity.

B. Short Answers

1. Proteomics

2. the Bohr effect

3. the α-helix

4. extremes of pH; the use of 8 M urea

5. a random coil

6. oligomeric proteins

7. the Fe^{2+} ion of the heme group

8. multienzyme complex

9. renaturation

10. positive cooperativity of binding

11. 2,3-*bis*phosphoglycerate (BPG)

12. Molecular chaperones

13. right-handed helix (α-helix or 3_{10} helix)

14. antigens; antibodies

15. allosteric modulators; allosteric proteins

C. Problems

1. The advance along the axis in an α-helix is about 0.15 nm per residue, so the segment is 0.15 nm/residue × 20 residues = 3.0 nm in length.

2. Pauling showed that there are about 3.6 residues per turn, so 20 residues × 1 turn/3.6 residues = 5.6 turns.

3. Carbonyl oxygens of peptide bonds are hydrogen bonded to the α-amino nitrogen of the fourth residue "ahead" (toward the C-terminus of the chain). Exceptions are that the first four α-amino nitrogens and the last four carbonyl oxygens of the α-helical segment are not involved in hydrogen bonding in the α-helix. The hydrogen bonds are roughly parallel to the helix axis.

4. Draw a linear protein segment of 10 residues. Label the side chains R_n, R_{n+1}, and so on through R_{n+9}. Draw hydrogen bonds between the carbonyl oxygen of residue n to the amide hydrogen of residue $n + 4$, as indicated in Section 4.4 of your text. This procedure will result in a variance with what is stated in your text because we have assumed that all atoms of the 10 residues are "in" the helix. With 10 residues, there are 10 carbonyl oxygens and 10 α-amino nitrogens. It takes one of each to make a hydrogen bond, so there is the potential for 10 hydrogen bonds. However, four α-amino nitrogens at the N-terminal end of the helix and four carbonyl oxygens at the C-terminal end of the helix are not involved in hydrogen bonding. Thus there are 10 – 4 = 6 hydrogen bonds. That represents 60% of the possible hydrogen bonds. **Note**: This percentage will change with the length of the chain. For a helical segment of 20 residues, there are 16 hydrogen bonds, representing 80% of possible hydrogen bonds. Does it appear reasonable that the number of hydrogen bonds in any α-helix will be n–4, where n is the number of residues in the helix?

5. The first three residues, if able to participate at all, will contribute less than one turn (3 residues × 1 turn/3.6 residues = 0.8 turns or 3 residues × 0.15 nm/residue = 0.45 nm). Proline is generally a helix breaker and will usually form a bend. The next six residues could form 6 residues × 1 turn/3.6 residues = 1.7 turns or 6 residues × 0.15 nm/residue = 0.90 nm of helix before the next proline bends the chain.

6. Normal collagen contains hydroxyproline and hydroxylysine residues that are formed by hydroxylation of proline and lysine, respectively. Proline is hydroxylated by the action of prolyl hydroxylase, which requires vitamin C for its activity. Without vitamin C, proper amounts of hydroxyproline and hydroxylysine are apparently not produced and collagen cannot form fibers properly. This must result in abnormal collagen that is unable to perform the normal structural functions.

7. Yes. Some lysine residues in collagen are converted to allysine by the action of lysyl oxidase. An allysine residue in one collagen chain can react with a lysine residue in another collagen chain to form a crosslink called a Schiff base. Also, two allysine residues in separate chains can react via an aldol condensation to form a different kind of crosslink. See Figure 4.38 in your text.

8. RNase A contains four disulfide bonds in its native, biologically active form that are broken by the action of the 2-mercaptoethanol. For the denatured, inactive protein to regain biological activity, it must regain its native conformation with the four disulfide bonds formed between the original (native) cysteine partners. In the presence of urea, the protein is prevented from properly refolding and reoxidation produces random disulfide bond formation for about 99% of the molecules. One may conclude that the protein chain must be able to fold into its native conformation, which can occur with both urea and 2-mercaptoethanol removed, so that the native disulfide bonds can reform and biological activity be regained.

9. Figure 4.50 in your text shows that hemoglobin is roughly 45% saturated (Y = .45) with oxygen at pH 7.2 when $pO_2 = 30$ torr. At pH 7.6, the degree of oxygen saturation is increased to about 73%. The ratio of oxygen saturation, then, is 0.73/0.45 = 1.6. Therefore, at pO_2 of 30 torr, a change of pH from 7.2 to 7.6 allows hemoglobin to hold 1.6 times as much oxygen.

10. Note that although these reactions are reversible, product (CO_2) formation is favored in the lungs where CO_2 concentration is low.

   (a) [insert figure]

   (b) [insert figure]

   (c) Since only four N-terminal amino groups per hemoglobin molecule are potentially available for carbamate adduct formation, it would seem that this might not be an important means of CO_2 transportation. Thomas Devlin, in *The Textbook of Biochemistry with Clinical Correlations*, 3rd Edition, 1992, page 1036, estimates the transport of carbon dioxide in its major forms as:

$$HCO_3^- \quad 78\%$$
$$\text{dissolved } CO_2 \quad 9\%$$
$$\text{carbamino hemoglobin} \quad 13\%$$

11. An immunoassay employs antigen-antibody reactions for the determination of chemical substances. The specificity of the antibody for a particular antigen (such as a protein) makes the assay for that antigen possible. The enzyme-linked immunosorbent assay, or ELISA, is used to detect small amounts of specific proteins. For example, it is the basis for a pregnancy test in which the placental hormone chorionic gonadotropin is detected in the urine of the female within a few days after conception. (Voet & Voet, *Biochemistry*, 1990, pages 77–78.)

12. It has been shown that small substrate molecules that diffuse into crystals of an enzyme are bound and transformed into product(s), indicating that the crystallized enzyme is present in its catalytic conformation.

13. A leucine zipper is a common structure in DNA-binding proteins that involves two segments of α-helix that are slightly coiled about each other. This coiled coil structure is maintained by hydrophobic interactions of the side chains of leucine and other hydrophobic residues.

D. Additional Problems

1. The ratio of the molar mass of the native protein to that of the monomer is: $8 \times 10^4/1.9 \times 10^4 = 4.2$ or 4 (to one significant figure). Since the native protein contains only one kind of protein chain, it must contain four chains of this monomer. Therefore, the native protein is a tetramer, specifically, a homogeneous tetramer.

2. A carbon-carbon bond requires about 345 kJ/mole to break (Holtzclaw, *General Chemistry*, 9th Edition, 1991, page 194). A hydrogen bond requires about 2–20 kJ/mol, or an average of 9 kJ/mol for its disruption (Figure 2.13 of your text.). Therefore, 345/9 = 38 hydrogen bonds to provide the energy equivalent of one carbon-carbon covalent bond.

3. (a) Six disulfide bridges involve 12 cysteine residues. Random reformation of the native disulfide bridges can be found by the probability calculation:

$$P = 1/11 \times 1/9 \times 1/7 \times 1/5 \times 1/3 \times 1/1 = 1/10,400$$

This probability is about 0.01%. The fact that there was 8% recovery suggests that something other than random chance played a role.

(b) Trypsin is synthesized in the body in the form of the inactive zymogen, trypsinogen. Active trypsin is formed by the removal of a segment of residues from the N-terminus of trypsinogen. Thus, the primary structure of the zymogen form, which determined the zymogen tertiary structure, is not the same as that remaining in trypsin. Hence, the original tertiary structure of trypsin might not be expected to be regained after denaturation.

4. Myoglobin contains 153 residues, 121 of which (Voet & Voet, page 219) are involved in the 8 regions of α-helix, so that the potential for 121 hydrogen bonds exists. Since each α-helix will have four α-amino nitrogens (and four carbonyl oxygens) that do not form peptide bonds, there will be $4 \times 8 = 32$ fewer hydrogen bonds than theoretically possible. So, the number of hydrogen bonds would be $121 - 32 = 89$. At 9 kJ/mol, there would be 89 hydrogen bonds × 9 kJ/mol/hydrogen bond= 801 kJ/mol stabilization.

5. The polypeptide chains in α-keratin are in the α-helix form. If you consider stretching out an α-helix until it is in the elongated β-sheet form, you can calculate the percent extension. The distance of one residue in an α-helix is 0.15 nm. In a pleated sheet, it is 0.32–0.34 nm. The percent of extension is then:

$$(0.33 - 0.15)\text{nm} \times 100/0.15 \text{ nm} = 120\%$$

6. For the dissociation:

$$MbO_2 \leftrightarrow Mb + O_2$$

The dissociation constant, K_d, is:

$$K_d = \frac{[Mb]\, pO_2}{[MbO_2]}$$

At half-dissociation, [Mb] = [MbO2] and pO2 = P50. Placing these values in the dissociation constant expression gives: K_d = P50. From this,

$$P_{50} = \frac{[Mb]pO_2}{[MbO_2]} \text{ and,}$$

$$\frac{[Mb]}{[MbO_2]} = \frac{P_{50}}{PO_2} = \frac{2.8 \text{ torr}}{6 \text{ torr}} = \frac{2.8}{6}$$

$$\%[Mb] = \frac{[Mb]}{[Mb] + [MbO_2]} \times 100 = \frac{2.8 \times 100}{2.8 + 6} = 32\%$$

7. Myoglobin contains 153 residues, of which 89 may be involved in hydrogen bonds in α-helix regions. Denaturation requires 89 residues × 0.4 kJ/residue = 35.6 kJ. At 9 kJ/mol per hydrogen bond, the number of hydrogen bonds disrupted is 35.6 kJ × 1 H-bond/9 kJ = approximately 4 H-bonds! This is in concert with the statement in Section 4.9 in your text that indicates how few hydrogen bonds need to be disrupted to cause denaturation of a globular protein.

Chapter 5

A. True-False

1. <u>False</u>.   Some RNA molecules called ribozymes also possess catalytic activity.

2. <u>False</u>.   Many enzymes require no cofactors.

3. <u>True</u>.   Enzymes are usually rather specific in terms of the substrates they recognize. In many cases, the D-form of a sugar is recognized but not the L- form (its mirror image or enantiomer).

4. <u>False.</u>   The substrate must fit specifically in the active site of the enzyme.

5. <u>False</u>.   Complexes of substrate analogs and enzymes have been observed by x-ray crystallography.

6. <u>False.</u>   This is a description of a ping-pong mechanism.

7. <u>False.</u>   Neither can be precisely determined from such a plot. Use of a Lineweaver-Burk plot gives better approximations.

8. <u>False.</u>   The *lower* the K_m, the greater the affinity of the enzyme for its substrate.

9. <u>True.</u>   As a consequence, intracellular enzymatic rates are sensitive to small changes in substrate concentrations.

10. <u>True.</u>   As substrate binds to one subunit, a conformation change occurs that facilitates binding of more substrate to the other subunits.

11. <u>True.</u>   These enzymes have one or more modulator binding sites (perhaps for both positive and negative modulators) as well as the active site. Almost all regulatory enzymes discovered to date are oligomeric.

12. <u>True.</u>   Such action permanently disables the enzyme unless the inhibitor is chemically removed. See the discussion involving diisopropylfluorophosphate in Section 5.8 of your text.

B. Short Answers

1. active site
2. enzyme-substrate complex
3. isomerases
4. catalytic constant, kcat (or turnover number)
5. specificity constant; specificity of the enzyme for its substrate(s)
6. covalent modification
7. competitive
8. noncompetitive
9. uncompetitive
10. feedback
11. regulatory sites
12. activator; inhibitor
13. converter enzyme

C. Problems

1. (1) Enzymes function as catalysts; (2) they catalyze highly specific reactions; (3) they can couple reactions; (4) their activity can be regulated.

2. Coupled reactions are reactions that would normally occur separately, but are connected by use of the product of one acting as the substrate of another. *In vivo*, the hydrolysis of ATP may serve to provide energy to drive an energetically unfavorable reaction and to remove the ATP formed as a product of the reaction.

3. Figure 5.2 in your text illustrates the solution to this problem. More product is formed per unit time in response to increasing amounts of enzyme. The system should be buffered and the temperature kept constant at a pH and temperature that produce optimum reaction velocity.

4.

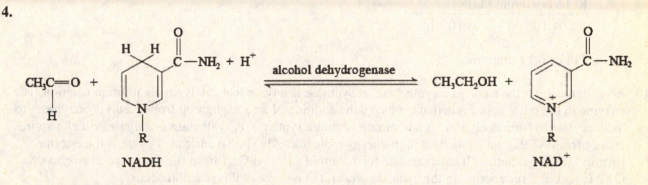

5. Since NADH absorbs light at 340 nm, changes in the NADH concentration can be estimated at this wavelength. As a starting material, NADH would be present in large quantity as the reaction begins. It is more difficult to detect small changes in a large amount of a substance than it is to detect the appearance of a substance (NADH) when none was previously present, as would be the case for the reverse reaction.

6. Using the Michaelis-Menten equation:

$$v_o = \frac{V_{max}[S]}{K_m + [S]}; \qquad \text{let } [S] = 2K_m. \text{ Then}$$

$$v_o = \frac{V_{max}(2K_m)}{K_m + (2K_m)}; \quad v_o = \frac{V_{max}(2K_m)}{3K_m};$$

$$v_o = 2/3\, V_{max}.$$

Similarly, for $[S] = 3K_m$, $v_o = 3/4\, V_{max}$. Also, for $[S] = 4K_m$, $v_o = 4/5\, V_{max}$.

7. Competitive inhibition is said to be overcome by an excess of substrate. Since the substrate and a competitive inhibitor compete with each other for the active site of a particular enzyme, a large excess of substrate insures that substrate molecules will bind to the active site nearly to the exclusion of competitive inhibitor molecules. For example, if the ratio of competitive inhibitor molecules to substrate molecules is 1/9, there is a 10% chance of inhibitor molecules binding to the active site. If the ratio is 1/99, there is only a 1% chance of inhibitor binding to the active site. The more substrate present, the less the likelihood of the inhibitor binding to the active site.

8. Measure the amount of product formed in a certain time period, at specific time intervals. The slope of the straight-line portion of the curve near time = zero gives the initial velocity, vo, as shown in text Figure 5.3.

9. $v_o = \frac{V_{max}[S]}{K_m + [S]};$ \qquad Solving for V_{max} gives:

$$V_{max} = \frac{v_o(K_m + [S])}{[S]}.$$ Using the data given:

$$V_{max} = \frac{0.16\,\mu mol/min(2.0 \times 10^{-5}\,M + 0.15\,M)}{0.15\,M}$$

$$V_{max} = \frac{0.024\,\mu mol/min\,M}{0.15\,M} = 0.16\,\mu mol/min$$

(This V_{max} indicates that the substrate concentration used (0.15 M) must have virtually saturated the enzyme. Thus, for [S] = 5.0 × 10⁻⁴,

$$v_o = \frac{(0.16\,\mu mol/min)(5.0 \times 10^{-4}\,M)}{2.0 \times 10^{-5}\,M + 5.0 \times 10^{-4}\,M};$$

$v_o = 0.114$ or 0.11 μmol/min.

10. A synthase is not the same as a synthetase. A synthase is a lyase that catalyzes an addition reaction. The enzyme in the citric acid cycle that catalyzes the addition of an acetyl group from acetyl coenzyme A to oxaloacetate to form coenzyme A and citrate is citrate synthase. A synthetase is a ligase, which requires energy from ATP or an equivalent high-energy molecule. Also in the citric acid cycle is the enzyme succinyl CoA synthetase. It catalyzes the formation of succinyl CoA from succinate and coenzyme A. GTP is required for energy. In the cycle, however, the *reverse* of this reaction occurs.

11. Such concentrations allow the enzyme to respond proportionally to changes in substrate concentrations in the cell and prevent the accumulation of substrate under changing metabolic conditions.

12. (a) From the velocity data given, the V_{max} appears to be 50.0 μmol/min. (b) Use that value, and calculate K_m from the data determined from any single measurement. For example, use measurement number 3, for which [S] = 1.0 × 10⁻⁴ and v_0 = 41.0 μmol/min. This data will also be used in part (c).

$$v_o = \frac{V_{max}[S]}{K_m + [S]};$$ Solve for K_m:

$$K_m = \frac{V_{max}[S]}{v_o} - [S];$$ Use V_{max} and the data from measurement number 3:

$$K_m = \frac{(50.0\,\mu mol/min)(1.0 \times 10^{-4}\,M)}{41.0\,\mu mol/min} - 1.0 \times 10^{-4}\,M$$
$K_m = 2.2 \times 10^{-5}$ M.

(c) A 30.0% decrease in the velocity (41.0 μmol/min) corresponds to a velocity of 28.7 μmol/min for [S] = 1.0 × 10⁻⁴ M. This gives an apparent K_m ($K_m{}^{app}$) of:

$$K_m{}^{app} = \frac{(50.0\,\mu mol/min)(1.0 \times 10^{-4}\,M)}{28.7\,\mu mol/min} - 1.0 \times 10^{-4}\,M$$

$K_m{}^{app} = 7.42 \times 10^{-5}$ M.

Use the relationship $K_m{}^{app} = K_m(1 + [I]/K_i)$:

$$Ki = \frac{K_m[I]}{K_m{}^{app} - K_m}.$$ Substitute the data from measurement number 3:

$$K_i = \frac{(2.2 \times 10^{-5} \text{ M})(1.4 \times 10^{-4} \text{ M})}{(7.42 \times 10^{-5} \text{ M}) - (2.2 \times 10^{-5} \text{ M})}$$

$K_i = 5.9 \times 10^{-5}$ M.

13. The category is hydrolases. Both enzymes catalyze hydrolysis. Trypsin catalyzes hydrolysis of peptide bonds and α-amylase catalyzes hydrolysis of glycosidic bonds in starch.

D. Additional Problems

1. Start with the Michaelis-Menten equation and take the reciprocal of both sides:

$$v_o = \frac{V_{max}[S]}{K_m + [S]}$$

$1/v_o = K_m + [S]/V_{max}[S]$; separate terms and cancel [S].

$1/v_o = K_m/V_{max}[S] + [S]/V_{max}[S]$

$1/v_o = K_m/V_{max}[S] + 1/V_{max}$ then,

$1/v_o = K_m/V_{max} \cdot 1/[S] + 1/V_{max}$.

Since this is in the form of the equation for a straight line, K_m/V_{max} is the slope of the line and $1/V_{max}$ is the intercept on the $1/v_o$ axis. Thus, the values of K_m and V_{max} can be evaluated graphically.

2. Start with the Michaelis-Menten equation and multiply both sides by $K_m + [S]$:

$$v_o = \frac{V_{max}[S]}{K_m + [S]}$$

$$v_o(K_m + [S]) = \frac{V_{max}[S](K_m + [S])}{K_m + [S]};$$ Expand and cancel terms:

$v_o K_m + v_o[S] = V_{max}[S];$ Subtract $v_0[S]$ from both sides:

$v_o K_m = V_{max}[S] - v_o[S];$ Divide both sides by K_m:

$$v_o = \frac{V_{max}[S] - v_o[S]}{K_m};$$ Collect terms and factor out [S]:

$$v_o = \frac{(V_{max} - v_o)[S]}{K_m};$$ Divide both sides by [S] and separate terms:

$$\frac{v_o}{[S]} = \frac{V_{max}}{K_m} - v_o\left(\frac{1}{K_m}\right).$$

This is a slightly rearranged form of the equation for a straight line. [In this case, it is y = b + (-xm)]. Therefore, a plot of $v_o/[S]$ vs. v_o gives a straight line of negative slope. Slope = $-1/K_m$. The intercept on the v_o axis = V_{max} and the intercept on the $v_o/[S]$ axis = V_{max}/K_m. Determination of these graphical points allows evaluation of the Michaelis constant and maximum velocity for the system.

3. The rate of 30. mmoles/12 min = 2.5 mmoles of product C formation per minute. An International Unit (I.U.) is defined as the number of micromoles of substrate transformed per minute. Therefore, the number of I.U. is:

$$\text{I.U.} = \frac{2.5 \text{ mmol C}}{\text{min}} \times \frac{10^3 \text{ } \mu\text{mol C}}{\text{mmol C}} \times \frac{2 \text{ } \mu\text{mol A}}{1 \text{ } \mu\text{mol C}}$$

I.U. $= 5.0 \times 10^3$ μmol A/min $= 5000$ I. U.

4. The specific activity of an enzyme solution is the number of International Units per mg of protein in the solution used.

Specific Activity $= 5000$ I.U./2.5×10^{-4} mg protein
Specific Activity $= 2.0 \times 10^7$ μmol A transformed per minute per milligram of protein in the solution.

5. The catalytic constant, k_{cat}, or turnover number, is the number of micromoles of substrate transformed per second per micromole of active sites. To determine the catalytic constant, convert the specific activity to a per second-basis and determine the number of micromoles of enzyme as shown:

$$k_{cat} = \frac{\left(5.0 \times 10^3 \ \mu\text{mol A/min}\right) \times 1\,\text{min}/60\,\text{sec}}{2.5 \times 10^{-4}\ \text{mg E} \times \dfrac{1\,\text{mmol E}}{12,500\,\text{mg E}} \times \dfrac{10^3\ \mu\text{mol E}}{1\,\text{mmol E}}}$$

$k_{cat} = 4.2 \times 10^6$ sec^{-1}.

Note that you cannot cancel micromoles of A by micromoles of E. The unit sec^{-1} actually means micromoles of substrate transformed per second per micromole of active sites. Remember that the enzyme in this problem was reported to have just one active site per molecule.

6. Since the data do not reveal V_{max}, determine it and K_m using a Lineweaver-Burk plot. First, put the kinetic data in reciprocal form as shown:

Measurement Number	1/[S], M^{-1}	1/v_o, min/μmol
1	5.88×10^5	0.100
2	2.63×10^5	0.050
3	8.33×10^4	0.022
4	4.35×10^4	0.017
5	1.18×10^4	0.012

Figure 5.1 shows the plot of this data as line D.4. From the graph, the y-intercept is 0.010 min/μmol. Since the y-intercept $= 1/V_{max}$,

$$V_{max} = 1/0.010 \text{ min/}\mu\text{mol} = 100 \ \mu\text{mol/min}.$$

The y-intercept is -6.7×10^4 M^{-1}. Therefore,

$$-6.7 \times 10^4 \text{ M}^{-1} = -1/K_m$$
$$K_m = 1/6.7 \times 10^4 \text{ M}^{-1} = 1.5 \times 10^{-5} \text{ M}.$$

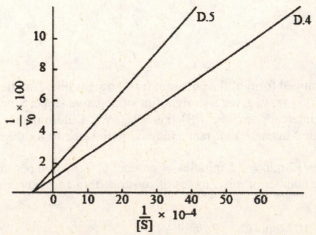

Figure 5.1   Lineweaver-Burk Plot for Problems D.4 and D.5

7. Put the data in reciprocal form and plot it on the same graph used in the previous problem. This gives a straight line (D.5) with the same x-intercept (approximately -6.7×10^4 M^{-1}) as shown in Figure 5.1. This indicates noncompetitive inhibition. The slope of this line is greater than that of line D.4. K_m for the system is unchanged as evidenced by the same x-intercept for both lines. Since the y-intercept of line D.5 is greater than that of line D.4, the V_{max} for the inhibited system is decreased. You may obtain values that are slightly different due to differences in graphing techniques or programs used. Apparently, noncompetitive inhibition is very rare. The data in this problem were arbitrarily chosen and are not from an actual experiment.

Chapter 6

A. True-False

1. <u>False</u>. It is a nucleophile. Having unshared electrons, it is attracted to an electron-deficient atom. It is named for that to which it is attracted. It is a nucleus lover.

2. <u>True</u>. The carbon atom carries a positive charge.

3. <u>True</u>.

4. <u>False</u>. Although the interior of a globular protein is hydrophobic, active site residues usually include several hydrophilic side chains. Such residues compose the *catalytic center* of the enzyme.

5. <u>True</u>. Covalent bonds form between a reactant and some enzyme active site group. One product is then released from the enzyme prior to release of the second product.

6. <u>False</u>. Moderately weak binding of reactants is necessary for efficient catalysis.

7. <u>False</u>. Entropy decreases as reactants are collected and oriented in a particular way, losing freedom of motion.

8. <u>True</u>. Such analogs are potent enzyme inhibitors.

9. <u>True</u>. Activation of chymotrypsinogen involves removal of two dipeptide segments of the zymogen chain leaving three peptide chains held together by a total of five disulfide bonds.

10. <u>True</u>. The catalytic triad consists of Asp, His, and Ser.

11. <u>True</u>. K_m values may be in the range of 10^{-6} to 10^{-5} M for coenzymes, but may be 10^{-3} to 10^{-2} M for substrates like urea or carbon dioxide. See text Section 6.5.

12. <u>True</u>. While some free radicals are involved in certain normal cellular chemistry, production of unwanted free radicals may lead to some undesirable reactions in the cell.

B. Short Answers

1. carbanion

2. mechanism

3. transition state

4. activation energy, Ea

5. diffusion-controlled

6. proximity effect

7. thermodynamic pit

8. low-barrier hydrogen bonds

9. serine

10. Transition-state stabilization

11. acid-base catalysis; covalent catalysis; proximity effects; transition-state stabilization

12. induced fit

13. histidine

14. Zymogens

C. Problems

1. (a) The carbonyl carbon atom has partial positive character and, therefore, would be attacked by the nucleophilic $Y:^-$ species.

   (b) The leaving group would be an anion. Extra electrons are brought to the carbon atom by the $Y:^-$ nucleophile so electrons must leave (e.g., as $X:^-$) to achieve a charge balance.

2. Intermediates are metastable species with lifetimes long enough (10^{-14}–10^{-13} sec, or longer) to allow them to be detected by sophisticated techniques. Transition states have even shorter lifetimes and have yet to be detected. The formation of an intermediate will involve a second activation energy barrier to be overcome before product is formed.

3. The species removed is H^-, a hydride ion. It is oxidation because the molecule from which the hydride ion is removed, the substrate, loses electrons.

4. (a) Ala + Cys–Lys–Met–Phe–Arg–Ala + Tyr–Gly

   (b) Ala–Cys–Lys + Met–Phe–Arg + Ala–Tyr–Gly

   (c) Ala–Cys–Lys–Met–Phe + Arg–Ala–Tyr + Gly

   (d)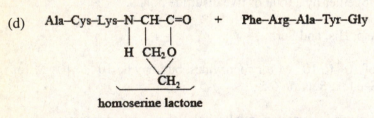

5. An enzyme lowers the overall activation energy by providing a multistep pathway in which the steps have lower activation energies than those of corresponding stages in the nonenzymic reaction.

6. In addition to the Asp–His–Ser catalytic triad, each enzyme has a binding site or specificity pocket that fits the side chain of the residue recognized. See text Figure 6.24.

7. Enteropeptidase is a duodenal enzyme that catalyzes removal of an N-terminal hexapeptide from trypsinogen to create active trypsin. Trypsin then activates other zymogens in the intestinal tract.

8. The bond cleaved that converts trypsinogen into trypsin is the Lys 6–Ile 7 peptide bond. Since trypsin catalyzes the cleavage of peptide bonds in which the carbonyl group is donated by lysine or arginine, the Lys 6 residue would be recognized by trypsin.

9. Although trypsin inhibitor contains residues that trypsin recognizes, the fit between the inhibitor and trypsin is very tight due partly to a complex network of hydrogen bonds. The fit is so tight that water cannot enter the active site to participate in hydrolysis.

10. This product was formed when DIFP reacted with an active site serine residue. Because the serine is involved in the catalytic mechanism, the enzyme was inactivated when the serine side chain was no longer available.

11. (a) The substrate binds in the active site with the carbonyl carbon of the scissile peptide bond close to the oxygen of the serine (Ser-195) in the catalytic triad.

(b) A proton is removed from the –OH of Ser-195 by attraction of the >N: of the His-57 side chain. This creates a more nucleophilic oxygen on Ser-195, which then attacks the carbonyl carbon of the scissile peptide bond.

(c) The carbonyl oxygen of the peptide bond becomes an oxyanion and moves into the *oxyanion hole* and is hydrogen bonded to -N-H groups of peptide bonds of Gly-193 and Ser-195.

(d) The proton abstracted by His-57 is donated to the nitrogen of the peptide bond, which causes its cleavage and the release of the first product, a free amine.

(e) Water enters the active site and donates a proton to His-57, resulting in formation of OH^-, which performs a nucleophilic attack on the carbonyl carbon of the serine ester group.

(f) His-57 donates the proton obtained from water to the Ser-195 oxygen, thus releasing the second product, a carboxylate ion, and regenerating the enzyme.

12. It is thought that the imidazolate form is the H^+ acceptor, rather than the neutral form of the side chain. See text Section 6.4.A.

13. The rate was greatly decreased. Although a $-COO^-$ group is present, it is farther away from the substrate, causing a lowered reaction rate.

D. Additional Problems

1. As you have learned, the maximum rate of reaction for an enzyme catalyzed system occurs with excess substrate (saturating levels). With excess substrate present, the rate does not change and zero order kinetic conditions with respect to substrate concentration are observed. In this case, the rate $= k[S]^0 = k$. Thus, the rates can be used in the Arrhenius equation.

$$\ln k = -E_a/R \times 1/T + \ln A$$

Write two equations using k_1 and T_1 and k_2 and T_2, respectively. Subtract the first equation from the second and solve for E_a.

$$\ln k_2 - \ln k_1 = E_a/R \times (1/T_1) - E_a/R \times (1/T_2)$$

Rearrange this to

$$\ln (k_2/k_1) = Ea/R(1/T_1 - 1/T_2), \text{ and solve for } E_a$$

$$E_a = \frac{R \times \ln k_2/k_1}{1/T_1 - 1/T_2}$$

$$= \frac{(8.314 \text{ J/mol K})(2.303 \log 36.6/25.2)}{(1/293 - 1/303) \text{ K}^{-1}}$$

$$= 2.7 \text{ J/mol or about } 27 \text{ kJ/mol}$$

2. The pH profile shown in Figure 6.1 indicates an acid-base mechanism involving the two aspartate residues in the active site. One aspartate has a pK_a of about 1.4, is therefore unprotonated at stomach pH, and thus acts as a proton acceptor (a general base). The other aspartate has a pK_a of about 4.3, is protonated at stomach pH, and acts as a proton donor (a general acid). Optimal activity occurs at the intermediate pH of about 2.7. The side chain of free aspartic acid has a pK_a of 3.9, so the pK_a values of both aspartate side chains are substantially altered by neighboring groups.

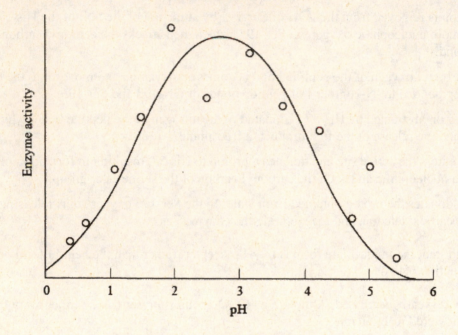

Figure 6.1   pH Profile of Pepsin

3. Figure 6.2 indicates that one product, P, appears before the other, indicating a multistep mechanism. This burst of the first product suggests that the substrate is cleaved with one product released rapidly. The remaining part of the substrate is likely covalently bound to a group in the active site of the enzyme and its release is controlled by the rate-limiting step of the mechanism.

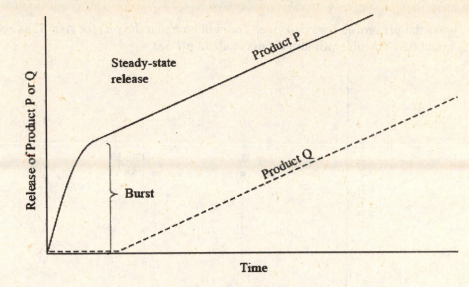

Figure 6.2   Burst Kinetics

4. For the nonenzymatic bimolecular reaction,

$$\text{rate} = k_2 \,[\text{A}][\text{B}]; \, k_2 = \frac{\text{rate}}{[\text{A}][\text{B}]}$$

From the data given,

$$k_2 = \frac{2.5 \times 10^{-2} \text{ mmol/2 min}}{[0.0100][0.0100]\text{M}^2};$$

$$k_2 = 1.25 \times 10^2 \text{ mmol/(min M}^2).$$

The enzyme-catalyzed reaction is unimolecular because reactants and catalytic groups are positioned near each other in the active site. Therefore,

rate = $_{k1}$[AB], where [AB] is the unimolecular reactant in the active site.

$$k_1 = \text{rate/[AB]} = \frac{4.0 \text{ mmol / 2 min}}{[0.0100]\text{M}}$$

$$k_1 = 2.0 \times 10^2 \text{ mmol/(min M)}$$

Since the effective molarity is the ratio of k_1/k_2,

$$\text{effective molarity} = \frac{2.0 \times 10^2 \text{ mmol min}^{-1} \text{ M}^{-1}}{1.25 \times 10^2 \text{ mmol min}^{-1} \text{ M}^{-2}};$$

Effective molarity = 1.6 M.

Note that the proximity effect is usually not the sole enzymatic catalytic mode.

5. Potential substrates with two, three, and four NAG units (A–B, A–B–C, and A–B–C–D) are rather stable to lysis in the presence of lysozyme. The highest rate of hydrolysis occurs with NAG$_6$ and the rate does not increase with additional NAG units. The specificity, therefore, is for the oligomer with six units in the active site (A–B–C–D–E–F) and all six sugar–binding sites must be occupied for maximal activity.

Since the A–B–C–D oligomer is not cleaved, none of these bonds (A–B, B–C, or C–D) are candidates.

Cleavage of NAG₅ suggests that the D–E bond is cleaved. The data indicate that lysozyme cleaves the bond indicated below in bacterial cell walls.

NAG–NAM–NAG–NAM–NAG–NAM
↑

6. Figure 6.3 shows the pH profile for lysozyme. You will note that the pK_a for Asp-52 is about 3.5 and that of Glu-35 is about 6.5. The pH optimum occurs at about pH 5.0.

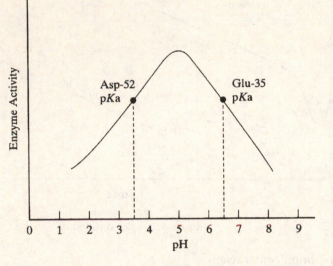

Figure 6.3   pH Profile of Lysozyme

Chapter 7

A. True-False

1. <u>False</u>.   The iron of iron sulfur clusters is not part of a heme group. It is, in fact, referred to as "non-heme iron." See text Figure 7.3.

2. <u>True</u>.   These are made from metabolites that occur normally in the organism.

3. <u>False</u>.   Water-soluble vitamins are readily excreted from the body and do not normally accumulate to toxic levels. Lipid vitamins (fat-soluble vitamins) can accumulate in the fatty tissues of the body and lead to hypervitaminosis.

4. <u>False</u>.   Pellagra is due to a deficiency of niacin required for synthesis of NAD+ and NADP+ coenzymes.

5. <u>True</u>.   The coenzyme form, thiamine pyrophosphate (TPP), is required by certain decarboxylase enzymes and by some transketolases.

6. <u>True</u>.   The anionic tail helps the coenzyme bind in the enzyme active site.

7. <u>False</u>.   Vitamin K is a fat-soluble vitamin, but it plays a role in proper synthesis of proteins involved in the blood-clotting mechanism. Vitamin E is a lipid vitamin that exhibits antioxidant activity.

8. <u>True</u>.   Coenzyme QH₂ is a weaker reducing agent than NADH. So, under standard conditions, the reaction will proceed as:

$$NADH + H^+ + Coenzyme\ Q \rightarrow NAD^+ + Coenzyme\ QH_2$$

9. <u>True</u>. The Fe^{3+} can accept an electron and become Fe^{2+}, which can donate an electron and return again to the Fe^{3+} state.

10. <u>False</u>. Vitamins A and K act as coenzymes, but D and E do not. Vitamin D is a generic descriptor of steroids that exhibit the biological activity of cholecalciferol, which includes a role in the absorption and deposition of calcium phosphate. Vitamin E is a generic descriptor for compounds that exhibit the biological activity of α-tocopherol, which includes antioxidant activity and a role in normal growth and fertility.

11. <u>False</u>. NADH absorbs at 340 nm, but NAD^+ does not.

12. <u>False</u>. Ribitol is a component of riboflavin, the vitamin required for production of FAD and FMN.

13. <u>True</u>. NAD^+ has an overall negative charge due to the phosphates and would migrate toward the positive electrode (the anode—so named for the kind of ions it attracts.)

B. Short Answers

1. coenzymes; essential inorganic ions

2. apoenzyme

3. holoenzyme

4. reagents with chemical properties unlike those of any of the amino acid side chains.

5. NAD^+ and $NADP^+$

6. reactive center

7. cosubstrate

8. prosthetic group

9. ylide (See text Figure 7.15.)

10. water-soluble; lipid-soluble

11. protein coenzymes; metal ions; iron-sulfur clusters; heme groups

12. glucosyl moiety

13. NAD and NADP

14. NADPH

C. Problems

1. Metal-activated enzymes are those that require metal ions such as K^+, Mg^{2+}, or Ca^{2+} for their activity. These ions may aid the binding of substrates to the enzyme. Metalloenzymes contain firmly-bound ions, such as those of iron, zinc, copper, and cobalt, in their active sites. These ions do not dissociate from the enzyme and usually participate in the reaction catalyzed.

2. Both types of coenzymes, cosubstrates and prosthetic groups, provide the active sites of enzymes with reactive groups not supplied by amino acid side chains in the active site of the enzyme.

3. *Coenzyme A* acts as a cosubstrate involving the transfer of acyl groups.

   NAD+ acts as a cosubstrate in oxidation-reduction reactions involving dehydrogenase enzymes.

   The vitamin *biotin* is a prosthetic group for ATP-dependent enzymes that catalyze carboxylation reactions.

   Thiamine pyrophosphate (TPP) is derived from vitamin B1 and is a prosthetic group for enzymes that catalyze decarboxylation or the transfer of two-carbon units containing a carbonyl group.

   Tetrahydrofolate is derived from the vitamin folate and serves as cosubstrate for enzymes that catalyze transfer of one-carbon groups including formyl, hydroxymethyl, and methyl groups.

   ATP is a metabolite coenzyme that acts as a cosubstrate for enzymes that catalyze the transfer primarily of phosphoryl or nucleotidyl groups.

FAD is derived from vitamin B$_2$, riboflavin, and serves as a cosubstrate for dehydrogenase enzymes that catalyze certain oxidation-reduction reactions.

4. Vitamin C exists as a lactone, which is an internal (cyclic) ester that forms when a carboxyl group reacts with a hydroxyl group on the same molecule. A molecule of water is eliminated and an ester, the lactone, results. Vitamin C is a required reducing agent in the hydroxylation of collagen.

5. The –SH group of the 2-mercaptoethylamine unit is the reactive center. It is involved in forming thioesters with acyl groups to be transferred.

6. The mechanism by which NAD$^+$ becomes reduced is by the acceptance of a hydride ion, H$^-$, which neutralizes the positive charge of NAD$^+$ and forms NADH. A proton is also removed from the substrate, but is released into solution. So, NADH and H$^+$ are products formed as the cosubstrate NAD$^+$ is reduced and the substrate is oxidized. Note that this is a two-electron oxidation-reduction.

7. It is an ordered sequential mechanism, not a ping-pong mechanism. Cosubstrate NAD$^+$ binds to the enzyme first, producing the holoenzyme, which then binds the substrate lactate. The product pyruvate is released first, followed by the release of NADH.

8. The substrate lactate is properly positioned by the formation of an ion pair with the positively charged side chain of Arg-171 and its own negatively charged carboxylate group. With lactate positioned, a proton is abstracted from the hydroxyl group on C-2 of lactate by the nearby side chain of His-195. As the electron pair that held the hydrogen of the hydroxyl group shifts back to the C-2 carbon of lactate, the hydrogen atom also on C-2 can leave with its electrons as a hydride ion and add to C-4 of the pyridine ring of NAD$^+$. Oxidation of the substrate is then complete and the product pyruvate is released. Note these details in text Figure 7.9.

9. FAD and FMN are held more tightly by their apoenzymes than is NAD$^+$ by its apoenzymes. FADH$_2$ and FMNH2 are rapidly oxidized in solutions containing oxygen, but are protected from this oxidation by remaining tightly bound in the active site of an enzyme. NADH does not have to be protected in this way since it is not susceptible to this kind of oxidation. Figure 7.1 shows the mechanisms by which NAD+ and FAD are reduced. Note that FAD is reduced by two one-electron transfers.

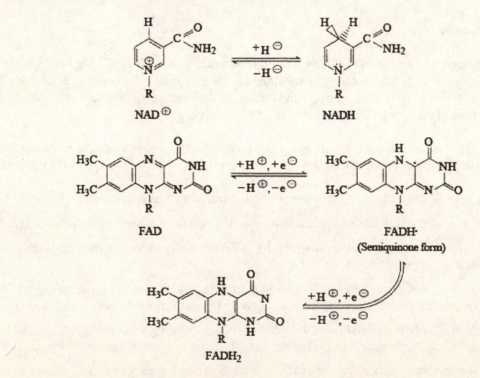

Figure 7.1   Mechanisms of Reduction of NAD$^+$ and FAD

10. Coenzyme Q (ubiquinone or Q) has a tail of 6–10 isoprenoid units that makes the coenzyme soluble in lipid membranes. Q is located in the inner mitochondrial membrane, where it participates in electron transport.

11. The reduction potential is a measure of the tendency to receive an electron from another substance. The cytochromes, even within a given class, have different reduction potentials due to different chemical environments (different amino-acid side chains) around the heme groups.

12. The Zn^{2+} ion promotes the ionization of a water molecule held in the active site. The hydroxide ion thus formed attacks the substrate CO_2 forming HCO_3^-, which is released as a substrate H_2O molecule adds.

13. S-Adenosylmethionine is the donor of virtually all the methyl groups used in biosynthetic reactions. (See text Section 7.3.)

14. FAD is yellow due to the system of conjugated double bonds in the isoalloxazine ring system. $FADH_2$ lacks the conjugated system of double bonds. FAD is the form in the bottle.

15. Humans obtain cobalamine from animals such as herbivores that obtain cobalamine from their intestinal bacteria. Mammalian liver stores the cobalamine and is therefore a good source of this micronutrient. Human liver can store a three- to five-year supply of cobalamine.

D. Additional Problems

1. Biotin becomes bound in the enzyme active site by forming an amide linkage with its carboxyl group and the ε-amino group of lysine. This forms the prosthetic group biocytin. Some texts indicate that ATP may react with bicarbonate to form carbonylphosphate, an active form of CO_2 that reacts with enzyme-bound biotin to form N-carboxybiotin. An active site base abstracts a proton from the β-carbon of pyruvate, forming a carbanion that performs a nucleophilic attack on N-carboxybiotin as indicated in Figure 7.2. Oxaloacetate is formed and the enzyme is regenerated.

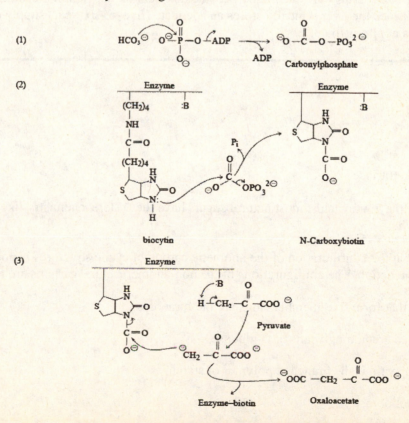

Figure 7.2 Mechanism of Pyruvate Carboxylase

2. Cancer involves uncontrolled cell division that requires DNA synthesis. DNA is composed of the nucleotides dAMP, dGMP, dTMP, and dCMP. The synthesis of dTMP, from dUMP, is catalyzed by the enzyme thymidylate synthase, which requires methylene tetrahydrofolate as the methyl group donor. In the reaction, dihydrofolate is produced and must be regenerated to tetrahydrofolate by the action of dihydrofolate reductase. If this enzyme is inhibited, supplies of tetrahydrofolate will be decreased, dTMP will decrease, and DNA synthesis and cell division will not occur. The antifolate drug methotrexate, an inhibitor of dihydrofolate reductase, has been used extensively in the treatment of cancer.

3. The cofactor that would be affected is biotin. Raw egg whites contain the protein avidin, which has a strong affinity for biotin and prevents absorption of biotin from the intestinal tract. Without sufficient biotin, reactions such as the production of oxaloacetate, required for initiation of the Krebs cycle, would not be possible and energy production would be impaired.

4. (a) Pyridoxal phosphate (PLP) is the coenzyme that forms Schiff bases during catalysis by enzymes such as transaminases. (b) An aldimine is a Schiff base formed from the condensation of a primary amine and an aldehyde. (c) An *internal aldimine* is formed by interaction of the aldehyde group of PLP with the ε-amino group of lysine in the active site of the enzyme as shown in text Figure 7.17. The *external aldimine* is formed by the interaction of the α-amino group of the substrate amino acid with the internal aldimine. The net result is a displacement of the active site lysine by the substrate amino acid. After hydrolysis of the intermediate, an α-keto acid product is formed as well as pyridoxamine phosphate (PMP). This form of the coenzyme serves as an amine group donor to an α-keto acid substrate to complete the transamination reaction.

5. (a) The disease may develop due to a deficiency of vitamin B_{12}. (b) The coenzyme forms are methylcobalamin, which participates in transfer of methyl groups, and 5'-deoxyadenosylcobalamin, which participates in intramolecular rearrangements. (c) The vitamin contains cobalt and a heme-like group called corrin. Humans require very small amounts (about 3 micrograms per day). To be absorbed, the vitamin must combine with a protein called *intrinsic factor* that is secreted in the stomach. In older people, sufficient absorption of vitamin B_{12} may not occur due to inadequate secretion of the intrinsic factor, even though there is an adequate supply of the vitamin in their diet. Since vitamin B_{12} is not made by plants, strict vegetarians may eventually become deficient in this vitamin because they are not eating meat products. Since the liver typically stores an adequate (five- to six-year) supply of vitamin B_{12}, onset of the disease is not immediate.

Chapter 8

A. True-False

1. <u>False</u>.   It is a ketotriose.

2. <u>False</u>.   Unlike the amino acids, most natural sugars have the D-stereochemistry like that of D-glyceraldehyde.

3. <u>True</u>.   Note that the configuration of the anomeric carbon of cyclic structures is not fixed, but changes in an equilibrium reaction. The configurations of the non-anomeric chiral carbons are fixed.

4. <u>True</u>.   Those that form five-member rings are the furanoses.

5. <u>True</u>.   All three are homoglycans of D-glucose.

6. <u>False</u>.   Amylopectin is the branched polymer of starch.

7. <u>True</u>.   So is cellobiose.

8. <u>False</u>.   All but sucrose are reducing sugars.

9. <u>False</u>.   It has α-(1→4) linkages like starch, but glycogen is more highly branched than amylopectin.

10. <u>True</u>.

11. <u>True</u>.   They differ only in the –OH orientation at C-4.

12. <u>False</u>.   D-Mannonic acid has a –COOH group formed from oxidation of the C-6 primary alcohol group (–CH₂OH).

13. <u>True</u>.   It could be formed by the reduction of D-glyceraldehyde.

14. <u>True</u>.   The Gram-positive bacteria have a thicker peptidoglycan cell wall than do Gram-negative bacteria. The thicker peptidoglycan cell wall retains the dye and thus is "stained."

B. Short Answers

1. oligosaccharide—specifically, a pentasaccharide

2. anomeric carbon

3. anomers

4. aglycone

5. limit dextrin

6. D-galactonic acid

7. chitin; N-acetyl-D-glucosamine

8. peptidoglycan

9. mucins

10. D-psicose (See text Figure 8.5.)

11. sugar alcohol

12. lactose; cellobiose (or cellobiose; maltose, but not lactose; maltose)

C. Problems

1.

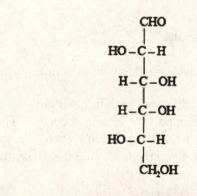

D-galactose   L-galactose (the enantiomer)

2.

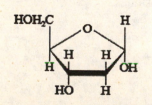

2-deoxy-α-D-ribofuranose

3. If 4% of the D-glucose residues are involved in branches, then 212 residues × 0.04 residues/branch = 8 branches. They occur about every 25 glucosyl residues.

4. The "head to head" linkage of D-glucose to D-fructose in the sucrose molecule ties up both anomeric carbons. Consequently, the anomeric carbons cannot open up to give the open chain form required for the reducing sugar reactions.

5. (a) D-glucose

   (b) *N*-acetyl-D-glucosamine

   (c) D-galactose and D-glucose

   (d) D-glucose

   (e) D-glucose

6.

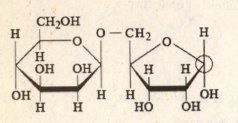

Yes, the ribose anomeric carbon (circled) is free to undergo ring opening.

7.

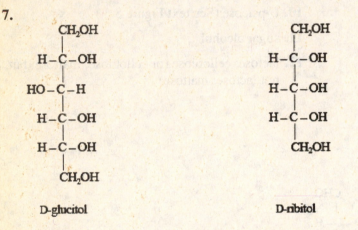

They cannot form cyclic structures since there is no aldehyde or ketone group available.

8. (a) The molar mass of glucose, $C_6H_{12}O_6$, is 180 g, but each internal glucosyl residue is less than 180 by the mass of a water molecule, eliminated when polymerization occurred. (This situation is just like that involving the mass of an amino acid residue in a protein chain.) Therefore, the residue mass of glucose is 180 − 18 = 162 g/mole. If the molar mass of glycogen is about 3.0×10^6 g, then the number of glucosyl residues is:

$$3.0 \times 10^6 \text{ g} \times 1 \text{ residue}/162 \text{ g} = 1.9 \times 10^4 \text{ residues.}$$

(b) Your text indicates that branch points occur in glycogen every 8 to 12 residues (an average of about 10%). Therefore, the number of glucosyl residues at branch points is:

$$1.9 \times 10^4 \text{ residues} \times 1 \text{ branch}/10 \text{ residues} = 1900 \text{ residues.}$$

9. The moles of glycogen in one pound:

$$454 \text{ g/lb} \times 1 \text{ mole}/3.0 \times 10^6 \text{ g} = 1.5 \times 10^{-4} \text{ mole/lb.}$$

The molecules of glycogen will be:

$$1.5 \times 10^{-4} \text{ mole} \times 6.02 \times 10^{23} \text{ molecules/mole} = 9.0\ 0215\ 10^{19} \text{ molecules.}$$

10. (a) There are four chiral centers. (b) Therefore, there should be 2^4 or 16 stereoisomers. (c) Text Figure 8.3 shows only eight aldohexoses, but there are an equal number of the L- series aldohexoses (the enantiomers of the eight D-aldohexoses).

11. There are five chiral centers in the Haworth or cyclic structures of the aldohexoses. Therefore, there can be 2^5 or 32 stereoisomers. The extra forms are due to the two anomeric forms (α and β) possible for each of the 16 stereoisomers of the previous problem.

12. It is a glycoside, specifically a galactoside. It is neither reducing nor shows mutarotation because the anomeric carbon cannot open up.

13. Only sucrose cannot form anomers because the C-1 of the glucose unit and the C-2 of the fructose unit are covalently fixed in position and are, therefore, not free to open up to form anomers.

14. Both molecules have only one reducing end.

D. Additional Problems

1. Your text indicates that, in aqueous solution, 36% of the glucose molecules are in the α-anomeric form and nearly all the rest, 64% are in the β-anomeric form. Using these quantities and the specific rotation values of each, calculate the equilibrium rotation that will eventually be reached in the aqueous solution due to mutarotation.

$$112.2°(0.36) + 18.7°(0.64) = 52°$$

(This neglects the small amount (less than 0.1%) of open chain form that is present.)

2. Since $[\alpha] = A/(l)(c)$, rearrangement to solve for c gives:

$$c = A/[\alpha](l) = 46.8°/(66.5°)(2.5) = 0.28 \text{ g/mL}$$

3. $151°(\alpha) + 53°(\beta) = 80°$; Also, $\alpha + \beta = 1$; Then, $\alpha = 1 - \beta$. Using this in the first equation gives,

$151°(1 - \beta) + 53°(\beta) = 80°$. Expanding gives,
$151° - 151\beta + 53°(\beta) = 80°$. Collection of terms gives
$-98° \beta = -71°$, then,
$\beta = .72$.
$\alpha = 1 - .72 = .28$.

Therefore, in the mixture of anomeric forms, there is 72% β and 28% α.

4. (a) The length of this segment of amylose helix is:

$$(900 \text{ residues})(1 \text{ turn/6 residues})(0.8 \text{ nm/turn}) = 120 \text{ nm}.$$

(b) Amylose forms a left-handed helix that is hydrated inside as well as outside, whereas the α-helix of proteins is right-handed and contains no water molecules in its interior. The α-helix of proteins contains fewer residues (3.6) per turn than does the amylose helix and has more turns in a given linear distance. If it were possible to have a polypeptide chain of 900 amino acid residues all in an α-helix, the length would be:

$$0.15 \text{ nm/residue} \times 900 \text{ residues} = 135 \text{ nm}.$$

5.

(a)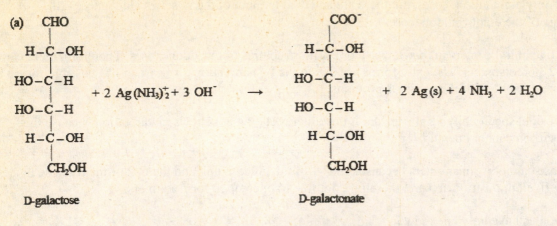

$$+ \, 2 \, Ag(NH_3)_2^+ + 3 \, OH^- \longrightarrow \quad + \, 2 \, Ag\,(s) + 4 \, NH_3 + 2 \, H_2O$$

D-galactose   D-galactonate

(Note: The two silver diammine complexes contain four ammonia molecules that are released when silver ion is reduced to metallic silver. Two water molecules are formed when two protons react with two hydroxide ions. One of the protons is released from the aldehyde group of the sugar after nucleophilic attack by hydroxide ion on the carbonyl carbon. The other proton comes from the carboxylic acid group formed in this reaction.

(b) The open chain form is in equilibrium with the cyclic anomeric forms. As some open chain form is irreversibly consumed, more will form as a new equilibrium is established. Thus, eventually, all of the sugar will be oxidized.

6. (a) Figure 8.1 shows the formation of a Schiff base between the primary amino group of a protein and the aldehyde group of a sugar (D-glucose, in this example). Formation of the Amadori product occurs as the result of two tautomerizations involving the Schiff base.

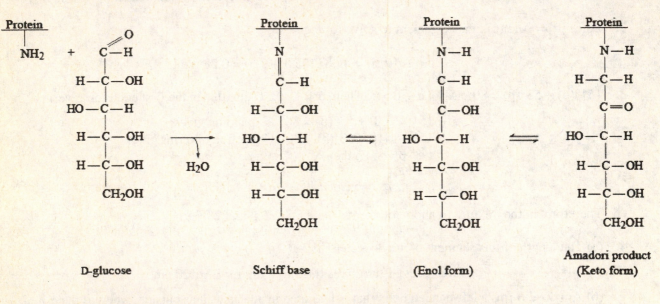

Figure 8.1   Formation of an Amadori Product from a Sugar and a Protein

(b) Figure 8.2 shows the formation of the cross link between proteins resulting from the reaction of an Amadori product, a glucose molecule, and the primary amino group of a protein. Seven molecules of water are eliminated in this multistep condensation.

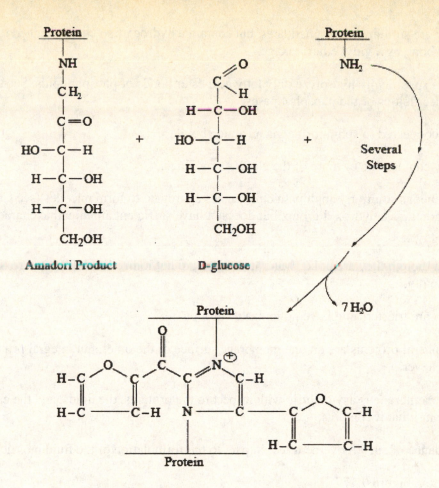

Figure 8.2 Formation of Cross-Linked Proteins

7. There are 20 possible disaccharides.

(a) There are four possible disaccharides involving linkages of the anomeric carbons of both sugars. They are: αG and βF (sucrose), αG and αF, βG and αF, and βG and βF.

(b) There are four possible disaccharides involving glucose in the alpha orientation and hydroxyl groups 1, 3, 4, and 6 of fructose.

(c) There are an additional four possible disaccharides involving glucose in the beta orientation and hydroxyl groups 1, 3, 4, and 6 of fructose.

(d) There are four possible disaccharides involving fructose in the alpha orientation and hydroxyl groups 2, 3, 4, and 6 of glucose.

(e) There are four possible disaccharides involving fructose in the beta orientation and hydroxyl groups 2, 3, 4, and 6 of glucose. Want to draw all of these?

Chapter 9

A. True-False

1. <u>False</u>.   The double bonds in polyunsaturated fatty acids are separated by a methylene group and are, therefore, not conjugated. Resonance cannot occur.

2. <u>True</u>.   Phosphatidyl choline (lecithin) is a glycerophospholipid, an amphipathic molecule.

3. <u>False</u>.   Although glycerophospholipids and most sphingolipids are ionic, cholesterol (a sterol) is uncharged.

4. <u>False</u>.   They are similar to phosphatidates, but contain a hydrocarbon chain attached to carbon-1 of the glyceryl backbone by a vinyl ether linkage.

5. <u>False</u>.   Some specifically hydrolyze only fatty acyl ester bonds of phospholipids. See text Figure 9.8 for hydrolytic specificities of the phospholipases.

6. <u>False</u>.   It is connected to sphingosine via an amide bond.

7. <u>True</u>.   They contain one or more of the NeuNAc moieties.

8. <u>False</u>.   In aqueous solution, amphipathic molecules aggregate to form micelles (see Chapter 2). Cholesterol contains a hydroxyl group, but does not have sufficient amphipathic character to form micelles.

9. <u>True</u>.   Even though they are polar, water molecules are not ionized and diffuse across membranes extremely rapidly.

10. <u>True</u>.   They are triesters and have no ionizable groups.

11. <u>False</u>.   Peripheral proteins are on one membrane surface or the other, but integral proteins are embedded in the lipid bilayer.

12. <u>True</u>.   The membrane is asymmetric with respect to the proteins, the lipids, and the carbohydrate moieties (if any) that it contains.

13. <u>True</u>.   Evidence of such movement led, in part, to the formulation of the fluid mosaic model.

14. <u>True</u>.   See text Figure 9.17.

B. Short Answers

1. Linoleate; linolenate

2. glycerophospholipids; sphingolipids; cholesterol

3. phosphate

4. serine; choline; ethanolamine

5. Gangliosides

6. Cerebrosides (or galactocerebrosides)

7. phospholipase D (See text Figure 9.8.)

8. Cerebrosides

9. Flippase (See text Section 9.9.)

10. liquid-crystalline phase; gel phase

11. flip-flop (or transverse diffusion)

12. peripheral; integral

13. G proteins

C. Problems

1. Energy storage, thermal insulation and padding (in animals), surface protection (waxes in plant cell walls), regulation of metabolism (in animals), and participation in cellular recognition are some of the diverse functions of lipids.

2. (a) Glucocerebroside yields sphingosine, a fatty acid, and glucose.
   (b) Sphingomyelin yields sphingosine, a fatty acid, a phosphate, and choline.
   (c) Lecithin yields glycerol, phosphate, choline, and two fatty acids.

3. Cholesterol molecules associate with the nonpolar tails of the phospholipids such that the hydroxyl group of cholesterol is at the aqueous surface of the lipid layer. The nearly planar cholesterol molecule lies approximately parallel to the fatty acid chains of surrounding lipids.

4. Eicosanoids are derivatives of polyunsaturated, 20-carbon fatty acids such as arachidonic acid (20:4). Prostaglandins are one type of eicosanoid. Prostaglandin E_2 constricts blood vessels. Thromboxane A_2 is involved in blood clotting. Leukotriene D_4 mediates smooth muscle contraction. (See text Figure 9.18 for structures.)

5. The hydrophobic character of triacylglycerols causes them to exclude water molecules as they aggregate. Since they are therefore not solvated by water, less space is required and more of these energy storage molecules can be stored in a given space.

6. The change in entropy of the water molecules provides the driving force. As water molecules orient themselves around a mono- or bilayer of lipids, the system becomes more ordered. Stabilization also occurs due to hydrophobic interactions of the nonpolar lipid tails.

7. The inner mitochondrial membrane and the plasma membrane of red blood cells are both rich in protein.

8. Carbohydrates are always found on the exterior surface of the cell membrane. They may be attached either to lipid molecules or to proteins.

9. Apparently Ca^{2+} and Mg^{2+} ions help to hold the peripheral proteins to the membrane surface via charge-charge interactions. Chelating agents complex these ions and disrupt the interactions between the peripheral proteins and the membrane. The peripheral proteins are thus freed from the membrane.

10. Primary active transport requires some direct source of energy such as light, ATP, or electron transport. The text example is that of bacteriorhodopsin, which uses light energy to generate a proton concentration gradient. Secondary active transport is driven by an ion concentration gradient. For example, the flow of protons into an *E. coli* cell down a concentration gradient allows lactose transport into the cell *against* its concentration gradient.

11. Most integral proteins appear to have one or more segments of nonpolar amino acid residues that span the nonpolar center of the lipid bilayer. Other areas of the protein are located on or near the membrane surfaces.

12. The triacylglycerol is quite hydrophobic with no charges or polar groups and is, therefore, most soluble in the chloroform component of the solvent. (The nonpolar solvents are the "moving phase" on silica gel.) Lecithin is a phosphatidate, which has a charged polar "head" region and is less soluble in the moving nonpolar phase and does not, therefore, travel very far from the origin.

13. Unsaturated fatty acid chains have *cis* double bonds that produce a bend in the chain and prevent close packing of the fatty acyl chains, thus providing a more fluid environment. Also, double bonds are a bit shorter than single bonds and that causes a little less interaction between the unsaturated hydrocarbon chains due to the shorter length. Cholesterol broadens the phase-transition temperature range of the bilayer. (See text Section 9.9.) It intercalates between lipid molecules, which causes decreased fluidity due to restricted mobility of adjacent fatty acyl chains. It increases fluidity by disrupting the ordered packing of fatty acyl chains of the lipid molecules. The net effect of cholesterol in membranes is that it helps maintain fairly constant fluidity despite fluctuations in temperature or degree of fatty acid saturation.

14. Endocytosis: A macromolecule binds to a protein receptor on the extracellular surface of the cell membrane. The membrane then invaginates, forming a lipid vesicle from part of the membrane. Exocytosis: Materials to be excreted from the cell are packaged in lipid vesicles that bud off from the Golgi apparatus. These vesicles fuse with the cell membrane and release their contents into the extracellular space.

15. Both are transmembrane proteins with a central passage for ions and small molecules. The term pore is used for bacteria and the term channel is used for animals.

D. Additional Problems

1. The number of molecules in one surface of the bilayer is:

$$2.86 \times 10^4 \text{ molecules/2 surfaces}$$
$$= 1.43 \times 10^4 \text{ molecules/surface.}$$

The surface area of one face is 100. μm^2 or 1.00×10^8 nm^2. The surface area per molecule is:

$$1.00 \times 10^8 \text{ nm}^2/\text{face} \div 1.43 \times 10^4 \text{ molecules/face}$$
$$= 6.99 \times 10^3 \text{ nm}^2/\text{molecule.}$$

2. Using 4.5 nm as the lipid bilayer thickness, as shown in Figure 9.1, the total contribution of the bilayer to the thickness of the liposome is:

$$2 \times 4.5 \text{ nm} = 9.0 \text{ nm.}$$

The diameter of the inner aqueous sphere is then $(40.0 - 9.0)$ nm = 31.0 nm. The volume of a sphere is $4/3 \ \pi r^3$. Thus, the volume is:

$$(4/3)(3.14)(15.5 \text{ nm})^3 = 1.56 \times 10^4 \text{ nm}^3.$$

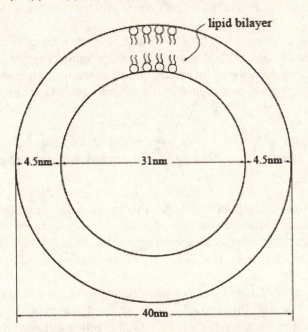

Figure 9.1   Cross-section of a Liposome

3. In an α-helix, each amino acid residue contributes 0.15 nm to the distance. Then,

$$4.5 \text{ nm} \times 1 \text{ residue/0.15 nm} = 30 \text{ amino acid residues.}$$

4. $s = (4Dt)^{1/2}$; $s^2 = 4Dt$, so $t = s^2/4D$. If $s = 2.0$ μm or 2.0×10^{-4} cm, then

$$t = (4.0 \times 10^{-8} \text{ cm}^2)/(4)(10^{-8} \text{ cm}^2/\text{sec}) = 1.0 \text{ sec.}$$

5. Lipids with a free –NH2 group, such as phosphatidyl serine and phosphatidylethanolamine, react with TNBS. Figure 9.2 illustrates the general reaction. Figure 9.3 illustrates the mechanism.

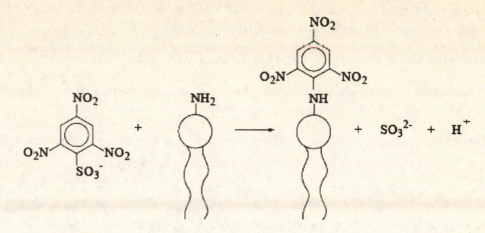

Figure 9.2   Reaction between TNBS and Lipids with a Free –NH₂ Group

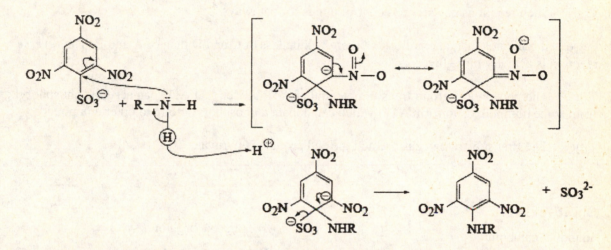

Figure 9.3   Mechanism for the Reaction of TNBS and Lipids with a Free –NH₂ Group

6. An α-helix incorporates 3.6 amino acid residues per turn and covers a distance of 0.54 nm. The average number of residues required to traverse the membrane in one direction is:

$$[4 \text{ nm} \times 1 \text{ turn}/0.54 \text{ nm}][3.6 \text{ residues/turn}] = 26.7 \text{ residues (per traversal)}.$$

Since 73% of the amino acid residues are in α-helix areas within the membrane, there are 247 residues × 0.73 = 180 residues in these areas. Therefore, the number of traversals is:

$$180 \text{ residues} \times 1 \text{ traversal}/26.7 \text{ residues} = 7 \text{ traversals}.$$

Chapter 10

A. True-False

1. <u>True</u>.   The two major divisions of metabolism are degradative (catabolic) and synthetic (anabolic).

2. <u>False</u>.   The flux of material through a pathway is controlled by availability of reactants, allosteric control, and covalent modification of interconvertible enzymes. Pathway flux can be increased or decreased as required by the cell.

3. <u>False</u>.   Energy transfer efficiency does *not* approach 100%. In general, it is less than 50%.

4. <u>True</u>.   Entropy is a measure of randomness or disorder. The NaCl is in a much less ordered state (greater randomness) in solution than in the very ordered crystalline state.

5. <u>True</u>.   The Gibbs free energy change (ΔG) has components of enthalpy and entropy.

6. <u>True</u>.   When ΔG is negative, the reaction can proceed without an external supply of energy. Note that the sign of ΔG, not $\Delta G^{\circ\prime}$, is the indicator of spontaneity.

7. <u>False</u>.   Other factors are also involved in determining whether a process is spontaneous. See the next question.

8. <u>False</u>.   The change in entropy, ΔS, is also a factor. The relationship is: $\Delta G = \Delta H - T\Delta S$.

9. <u>False</u>.   Cellular reactions occur at a steady state in which substrate is being supplied at about the same rate that product is being removed.

10. <u>True</u>.   Such reactions can serve as control points for pathways.

11. <u>False</u>.   It indicates that, at equilibrium, there will be less of C and D than of A and B. The *reverse* reaction is actually more favored.

12. <u>True</u>.   The phosphoryl-group-transfer potential of ATP is such that ADP can accept a phosphoryl group from some compounds to form ATP, which can donate a phosphoryl group to other compounds.

13. <u>True</u>.   The first substrate is glucose and the end product is pyruvate.

B. Short Answers

1. anabolic; catabolic

2. linear; cyclic; spiral

3. flux

4. increase; release

5. glycolysis

6. phosphoanhydride bond

7. reducing agent; oxidized

8. ATP; GTP; reduced cofactors

9. phosphagens; phosphocreatine;

phosphoarginine

10. energy-rich compounds

11. product; feedback inhibition; substrate; feed-forward activation

12. allosteric modulation; covalent modification

13. reduction potential

14. chemoheterotroph

15. protein folding; lipid bilayer formation

C. Problems

1. There is a turnover of cellular components. Synthesis and degradation of cellular components are regulated so that a reasonably constant concentration of these components is maintained. So, the statements are not in conflict.

2. Pathways (a) and (c) are anabolic (they involve synthesis) and pathway (b) is catabolic (it involves breakdown of fuel molecules).

3. Yes, it is possible under appropriate nonstandard conditions. While the change in standard free energy, $\Delta G^{\circ\prime}$, may be positive, it is the free energy change, ΔG, for the nonstandard set of conditions that must be negative for the reaction to proceed spontaneously.

4. In the expression that relates ΔG and $\Delta G^{\circ\prime}$,

$$\Delta G = \Delta G^{\circ\prime} + RT \ln [C][D]/[A][B],$$

the ln term is the only factor that can change. If the mass action ratio, Q, is less than one, which is the case when concentrations of reactants are greater than concentrations of products, the ln term will be negative. Still, the negative term must be larger than the positive $\Delta G^{\circ\prime}$ value for ΔG to also be negative.

5. Use the relationship:

$$\Delta G^{\circ\prime} = -RT \ln K_{eq}, \text{ and solve for } \ln K_{eq}.$$
$$\ln K_{eq} = \Delta G^{\circ\prime}/(-RT)$$

From text Table 10.3, $\Delta G^{\circ\prime}$ for this reaction is -21 kJ/mol. Using the value from the text for R and the standard temperature of 25°C converted to Kelvins,

$$\ln K_{eq} = \frac{(-21\,\text{kJ/mol})(10^3\,\text{J/kJ})}{(-8.315\,\text{J K}^{-1}\,\text{mol}^{-1})(298\,\text{K})}$$
$$= 21 \times 10^3/2.48 \times 10^3 = 8.47$$
$$K_{eq} = e^{8.47} = 4758 = 4.8 \times 10^3$$

6. Characteristically, *phosphorylation* activates the interconvertible enzymes of catabolic pathways, but those in anabolic pathways are usually activated by *dephosphorylation*.

7. Compartmentation allows (1) separate pools of metabolites, (2) opposing metabolic pathways to operate simultaneously, (3) high local concentrations of metabolites, (4) coordinated regulation of multiple enzymes, and (5) site-specific regulation of metabolic processes by specialized tissues.

8. The formation of one ATP from ADP and P_i requires 30. kJ/mol, the same quantity liberated when ATP is hydrolyzed, as indicated in text Table 10.1. Therefore, 32 moles of ATP $\times$ 30. kJ/mol ATP = 960 kJ. The *in vitro* production of 686 kcal/mol $\times$ 4.184 kJ/kcal = 2870 kJ. The percentage of energy captured from the glucose is:

$$\% \text{ capture} = \frac{960\,\text{kJ} \times 100}{2870\,\text{kJ}} = 33.45 = 33\%$$

9. $\Delta G = \Delta G^{\circ\prime} + RT \ln Q$, where Q is the mass action ratio of products to reactants,

$$\frac{[\text{glucose}][P_i]}{[\text{glucose 6 - phosphate}]} \quad {}_1$$

$$\Delta G = -14\,\text{kJ/mol} + (8.315 \times 10^{-3}\,\text{kJ/mol K})(310.\,\text{K}) \ln \frac{[2 \times 10^{-4}][5 \times 10^{-2}]}{[10^{-3}]}$$
$$= -14\,\text{kJ/mol} + (-11.87\,\text{kJ/mol})$$
$$= -25.87 = -26\,\text{kJ/mol}.$$

---

[1]The concentration of water as a reactant is not included in the ratio, Q, because it is essentially constant. The small amount used in the reaction is negligible compared with that available in the surrounding medium, the cell cytosol. In addition, the cellular water concentration is readily maintained, since plasma membranes are freely permeable to water.

10. (a) Assume the cell uses the energy supplied by the hydrolysis of ATP to AMP and PP_i to drive the reaction. This might be accomplished by using ATP to donate the AMP group to the carboxylic acid and then using the activated acid to react with the thiol.

$$R\text{–}COO^- + ATP \rightarrow R\text{–}CO\text{–}AMP + PP_i$$
$$R\text{–}CO\text{–}AMP + R'\text{–}SH \rightarrow R\text{–}CO\text{–}SR' + AMP$$

The sum of these reactions is:

$$R\text{–}COO^- + ATP + R'\text{–}SH \rightarrow R\text{–}CO\text{–}SR' + AMP + PP_i$$

Notice that this is the same net reaction obtained by adding the following processes for which the standard free energies are known:

$$R\text{–}COO^- + R'\text{–}SH \rightarrow R\text{–}CO\text{–}SR' + H_2O \quad \Delta G^{\circ\prime} = 10. \text{ kJ/mol}$$
$$ATP + H_2O \rightarrow AMP + PP_i \quad \Delta G^{\circ\prime} = -32 \text{ kJ/mol}$$

(b) The net $\Delta G^{\circ\prime}$ for this combination is -32 kJ/mol + 10. kJ/mol = -22 kJ/mol. Since this standard free energy change is negative, the reaction should proceed spontaneously. An additional driving force might be the hydrolysis of PP_i to 2 P_i, catalyzed by pyrophosphatase. The $\Delta G^{\circ\prime}$ for this reaction is -34 kJ/mol, which, added to the previous total, gives an overall total of -56 kJ/mol. (Note: this assumes that the appropriate enzymes for catalysis of each step are present in the cell.)

11. Calculate the standard free energy change from the relationship: $\Delta G^{\circ\prime} = -nF\Delta E^{\circ\prime}$. Use the standard reduction potentials from text Table 10.4 to calculate the change in standard reduction potentials, $\Delta E^{\circ\prime}$, using this relationship:

$$\Delta E^{\circ\prime} = E^{\circ\prime}_{red} - E^{\circ\prime}_{ox}$$

(Note that $_{red}$ and $_{ox}$ are used to indicate the half-reaction involving reduction and oxidation, respectively.) From text Table 10.4,

$$Q + 2 H^+ + 2 e^- \rightarrow QH_2 \qquad E^{\circ\prime} = 0.04 \text{ V}$$
$$\text{fumarate} + 2 H^+ + 2 e^- \rightarrow \text{succinate} \qquad E^{\circ\prime} = 0.03 \text{ V}$$
For the reaction,
$$\text{succinate} + Q \rightarrow \text{fumarate} + QH_2,$$
$$\Delta E^{\circ\prime} = 0.04 \text{ V} - 0.03 \text{ V} = 0.01 \text{ V}.$$

Use this change in standard reduction potentials in the equation given above.

$$\Delta G^{\circ\prime} = -(2)(96.48 \text{ kJ/V mol})(0.01 \text{ V})$$
$$= -1.9296 = -2 \text{ kJ/mol}.$$

12. Combination of the two steps yields a negative $\Delta G^{\circ\prime}$ (-3.0 kJ/mol) for the overall process, indicating that the process is energy-releasing and, therefore, spontaneous.

13. From text Table 10.4,

$$\text{pyruvate} + 2H^+ + 2e^- \rightarrow \text{lactate} \qquad E^{\circ\prime} = 0.18V$$
$$NAD^+ + 2H^+ + 2e^- \rightarrow NADH + H^+ \qquad E^{\circ\prime} = -0.32V$$

For the reaction $\quad$ pyruvate + NADH + $H^+ \rightarrow$ lactate + NAD^+

$$\Delta E^{\circ\prime} = 0.18V + 0.32V = 0.50V$$
$$\text{Then, } \Delta G^{\circ\prime} = -nF\Delta E^{\circ\prime} = -(2)(96.48 \text{kJ/V mol})(0.50V) = -96 \text{ kJ/mol}$$

D. Additional Problems

1. (a) Use the Henderson-Hasselbalch equation:

$$pH = pK_a + \log [A^-]/[HA]$$
where $A^- = ATP^{4-}$, and $HA = HATP^{3-}$.
$$\log [A^-]/[HA] = pH - pK_a,$$
$$\log [ATP^{4-}]/[HATP^{3-}] = 7.40 - 6.95 = 0.45.$$
$$[ATP^{4-}]/[HATP^{3-}] = 10^{0.45} = 2.8 \text{ or a ratio of 2.8 to 1.}$$
If the ratio is 2.8/1, the percentage of ATP^{4-} is:
$$\% \ ATP^{4-} = [2.8/(2.8 + 1)] \times 100 = 73.68 = 74\%.$$

(b) Doing the same calculation using pH 7.0 rather than 7.40 gives a percentage of 52% for the ATP^{4-} form.

2. (a) For the reaction:

$$\text{Glucose} + \text{ATP} \rightarrow \text{Glucose 6-phosphate} + \text{ADP},$$

the change in standard free energy is related to the equilibrium constant by:

$$\Delta G^{\circ\prime} = -RT \ln K_{eq}.$$
$$\ln K_{eq} = \Delta G^{\circ\prime}/-RT$$

$$\ln K_{eq} = \frac{-16.7 \text{ kJ/mol}}{(-8.315 \times 10^{-3} \text{ kJ/mol K})(310. \text{ K})}$$

$$\ln K_{eq} = \frac{16.7 \text{ kJ/mol}}{2.578 \text{ kJ/mol}} = 6.48$$

$$K_{eq} = e^{6.48} = 652.$$

This value and the negative standard free energy change indicate that the reaction should proceed spontaneously to the right.

(b) Use a series of approximations to determine the equilibrium concentrations. At equilibrium, [glucose 6-phosphate] = [ADP] = x, and [Glucose] = [ATP] = (0.0500 -x) M. The large K_{eq} value indicates that, at equilibrium, the concentrations of products are much greater than the concentrations of remaining reactants. Trial 1: If 90.% of reactants were consumed, x = (0.0500 M)(0.90) = 0.0450 M. The value of K_{eq} would be $(0.0450)^2/(0.00500)^2 = 81$, which is smaller than the known value of i_{eq}, indicating that greater than 90.% of the reactants must have reacted. Trial 2: If 98% of reactants were consumed, x = (0.0500 M)(0.98) = 0.0490 M. The value of K_{eq}, determined as in Trial 1 is 2,400. Since this is much larger than the calculated value of K_{eq}, less than 98% of the reactants reacted. Similar trials show that slightly more than 96.2% of the reactants were consumed. Thus, the concentrations of glucose 6-phosphate and ADP are about 0.0481 M and those of glucose and ATP are about 0.00190 M.

3.

$$\Delta G = \Delta G^{\circ\prime} + RT \ln \frac{[\text{glucose 6-phosphate}][\text{ADP}]}{[\text{glucose}][\text{ATP}]};$$

$$\ln \frac{[\text{glucose 6-phosphate}][\text{ADP}]}{[\text{glucose}][\text{ATP}]} = \frac{\Delta G - \Delta G^{\circ\prime}}{RT};$$

$$\ln \frac{[\text{glucose 6-phosphate}]}{[\text{glucose}]} + \ln \frac{[\text{ADP}]}{[\text{ATP}]} = \frac{\Delta G - \Delta G^{\circ\prime}}{RT};$$

$$\ln \frac{[\text{glucose 6-phosphate}]}{[\text{glucose}]} = \frac{\Delta G - \Delta G^{\circ\prime}}{RT} = \ln \frac{[\text{ATP}]}{[\text{ADP}]};$$

$$\ln \frac{[\text{glucose 6-phosphate}]}{[\text{glucose}]} = \frac{-33.9\,\text{kJ/mol} - (-16.7\,\text{kJ/mol})}{(8.315 \times 10^{-3}\,\text{kJ/mol K})(310.\,\text{K})} + \ln \frac{[13]}{1};$$

$$\ln \frac{[\text{glucose 6-phosphate}]}{[\text{glucose}]} = \frac{-17.2}{2.578} + 2.56 = -6.67 + 2.56 = -4.11;$$

$$\frac{[\text{glucose 6-phosphate}]}{[\text{glucose}]} = e^{-4.11} = 0.0164.$$

The ratio of glucose to glucose 6-phosphate is the reciprocal of 0.0164 or about 61 to 1.

4. $\Delta G = \Delta G^{\circ\prime} + RT \ln K_{eq};$

$$\ln K_{eq} = \frac{\Delta G - \Delta G^{\circ\prime}}{RT}$$

$$\ln \frac{[\text{ADP}][\text{P}_i]}{[\text{ATP}]} = \frac{\Delta G - \Delta G^{\circ\prime}}{RT}$$

$$\ln \frac{[\text{ADP}][\text{P}_i]}{[\text{ATP}]} = \frac{-41.8\,\text{kJ/mol} - (-30.\,\text{kJ/mol})}{(8.315 \times 10^{-3}\,\text{kJ/mol K})(310.\,\text{K})} = -4.58$$

$$\frac{[\text{ADP}][\text{P}_i]}{[\text{ATP}]} = e^{-4.58} = 0.0103.$$

Let $x = [\text{ADP}] = [\text{P}_i]$. When $[\text{ATP}] = 1$, $x = (1 \times 0.0103)^{1/2} = 0.101$. The ratio of ATP to ADP is then 1/0.101 or 9.9 to 1.

Since the ΔG for this reaction is even more negative than the standard free energy change, $\Delta G^{\circ\prime}$, the tendency for hydrolysis is even greater than at standard conditions.

5. (a) Both systems produce the same amount of stored energy as ATP (2 moles each), but glycolysis produces 2 NADH, which represents reducing equivalents capable of producing more ATP by their oxidation in the electron transport system.

(b) No net oxidation of glucose occurs when lactate is formed, since there are no reduced cofactors formed. Oxidation does occur when pyruvate is formed as noted by the 2 NADH formed.

(c) The system in which lactate is formed has the larger negative value of standard free energy change and thus has the greater tendency to proceed spontaneously as written.

Chapter 11

A. True-False

1. <u>False</u>. They are in the cytosol of eukaryotic cells.

2. <u>True</u>. The change in $\Delta G°'$ for the overall process is negative. Some of the energy derived from glycolytic catabolism of glucose is stored in ATP, formed in two pathway reactions.

3. <u>True</u>. Considering pyruvate the end product of glycolysis, oxidation is achieved without the *direct* use of oxygen.

4. <u>False</u>. A net of two moles of NADH is produced per mole of glucose. This occurs in step 6 in which glyceraldehyde-3-phosphate is oxidized to 1,3-*bis*phosphoglycerate.

5. <u>True</u>. Fructose 1,6-*bis*phosphate and 1,3-*bis*phosphoglycerate are the two intermediates with two phosphate groups.

6. <u>True</u>. Reactions 1, 3, 7, and 10 are catalyzed by kinases.

7. <u>True</u>. In yeast, pyruvate is decarboxylated to acetaldehyde, which is then reduced to ethanol.

8. <u>True</u>. The enzyme pyruvate decarboxylase is absent in humans.

9. <u>False</u>. This kind of intramolecular transfer is characteristic of enzymes in category 2, transferases.

10. <u>True</u>. Combination of the two stages results in a net production of two ATP per mole of glucose.

11. <u>True</u>. Entry of substrates as either glucose or as fructose 6-phosphate can be controlled by regulating these two reactions.

12. <u>False</u>. Only the D- isomer is formed. The enzyme is stereospecific.

13. <u>False</u>. Glucose is phosphorylated after entering the cell, which prevents it from passing back through the cell membrane.

14. <u>True</u>. One mole of either fructose, galactose, mannose, or glucose produces a net of two moles of ATP via glycolysis.

B. Short Answers

1. isozymes; phosphoenolpyruvate

2. phosphatases

3. fructose 1,6-*bis*phosphate; ATP

4. hexokinase; phosphofructokinase-1; pyruvate kinase

5. dihydroxyacetone phosphate

6. substrate-level phosphorylation

7. epimerase

8. the Pasteur effect

9. a deficiency of the enzyme lactase

10. citrate; ATP

11. pyruvate and ATP

12. 20

13. galactose

C. Problems

1. D-glucose + 2 ADP + 2 P_i + 2 NAD$^+$ → 2 pyruvate + 2 ATP + 2 NADH + 2 H$^+$ + 2 H_2O

2. Glycolysis would not proceed under these conditions because ATP is required in the first half of glycolysis at two separate steps. (Which steps are they?)

3. Oxidation of NADH by the electron transport system results in the formation of "several" ATP molecules.

4. Lactate is not constantly formed in mammalian cells. In most tissues, pyruvate is normally decarboxylated to yield acetyl CoA that fuels the Krebs cycle, under aerobic conditions. In tissues such as those of exercising muscle, in which oxygen is in low supply, some pyruvate is converted to lactate to keep glycolysis going by regenerating NAD$^+$ (required for which step?).

5.

Hexose	Initial Glycolytic Intermediate	Number of moles of Intermediate
mannose	fructose 6-phosphate	1
fructose	glyceraldehyde 3-phosphate and	1
	dihydroxyacetone phosphate	1
galactose	glucose 6-phosphate	1

6. If not for triose phosphate isomerase, only one half of the glucose molecule, D-glyceraldehyde 3-phosphate, would be utilized.

7. Glyceraldehyde-3-phosphate + NAD$^+$ + P_i + 2ADP → pyruvate + 2ATP + NADH + H$^+$

8. Figure 11.2 shows the regeneration of NAD$^+$ in muscle cells contrasted with that in yeast cells.

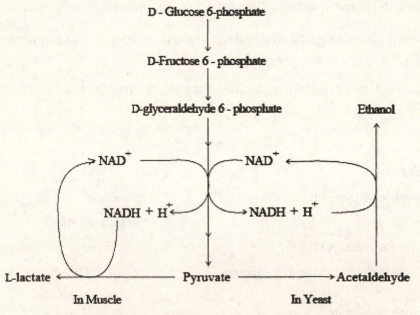

Figure 11.2 Regeneration of NAD$^+$ in Muscle and Yeast Cells

9. Enolase catalyzes the conversion of 2-phosphoglycerate to phosphoenolpyruvate. In the presence of an inhibitor, the substrate 2-phosphoglycerate would accumulate, as would other intermediates formed in previous, near-equilibrium reactions in the pathway.

10. Dihydroxyacetone phosphate would accumulate and present a problem unless disposed of by some other mechanism. Also, the individual would obtain only half the energy from glucose that a normal individual would obtain, since only the glyceraldehyde 3-phosphate half of the cleaved fructose 1,6-bisphosphate molecule would be available for energy production.

11. Figure 11.3 illustrates the reaction between iodoacetate and the cysteine sulfhydryl group of the enzyme. This is irreversible inhibition.

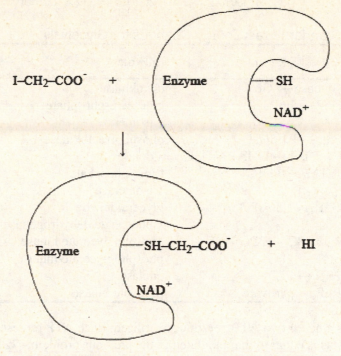

Figure 11.3 Iodoacetate Inhibition of Glyceraldehyde 3-Phosphate Dehydrogenase

12. ATP supplies phosphoryl groups in two reactions in the hexose stage. Inorganic phosphate, Pi, is the source of the other phosphoryl groups. One mole of P_i is used per mole of glyceraldehyde 3-phosphate.

13. Arsenate can cleave the enzyme-bound intermediate of the conversion of glyceraldehyde-3-phosphate to 1,3-bisphosphoglycerate. 1-Arseno-3-phosphoglycerate is formed instead. Although this product is converted to 3-phosphoglycerate, the energy-producing step involving 1,3-*bis*phosphoglycerate does not occur. This results in no net ATP being formed from glycolysis.

14. Pyruvate can be (a) converted to acetyl CoA, (b) carboxylated to form oxaloacetate, (c) reduced to ethanol in some species, and (d) reduced to lactate in some species.

D. Additional Problems

1. Use Table 11.1 and combine the standard free energy changes for reactions 1, 2, 3, 4, and 5. To that value, combine two times the standard free energy changes for reactions 6 through 10, since these reactions occur twice in consuming the two halves of the original glucose molecule.

$$\Delta G^{\circ\prime} = [-16.7 + 1.67 + (-14.2) + 24.0 + 7.66] + 2[6.28 + (-18.8) + 4.44 + 1.84 + (-31.4)] \text{ kJ/mol} = 2.4 + 2(-37.6) = -72.8 \text{ kJ/mol}$$

Table 11.1 Standard Free Energy Changes for Glycolysis

Step No.	Reaction	Enzyme	$\Delta G^{\circ\prime}$, kJ/mol
1	Glucose → G6P	hexokinase	-16.7
2	G6P → F6P	glucose 6-phosphate isomerase	+1.67
3	F6P → F1,6bisP	phosphofructokinase-1	-14.2
4	F1,6bisP → DHAP + G3P	aldolase	+24.0
5	DHAP → G3P	triose phosphate isomerase	+7.66
6	G3P → 1,3bisPG	glyceraldehyde 3-phosphate dehydrogenase	+6.28
7	1,3bisPG → 3PG	phosphoglycerate kinase	-18.8
8	3PG → 2PG	phosphoglycerate mutase	+4.44
9	2PG → PEP	enolase	+1.84
10	PEP → pyruvate	pyruvate kinase	-31.4

2. Glycolysis produces a net of two ATP per glucose utilized. Each ATP represents -30. kJ/mol, so the total energy from glycolysis is equal to that tabulated in the previous problem, -72.8 kJ/mol, plus that captured as ATP, $2 \times (-30. \text{ kJ/mol})$.

Total energy = -72.8 kJ/mol + (-60. kJ/mol) = -132.8 = -133 kJ/mol.

$$\text{The percent efficiency of energy capture} = \frac{-60 \text{ kJ/mol} \times 100}{-133 \text{ kJ/mol}} = 45\%$$

3. Formation of two lactates from the two pyruvates formed from one glucose represents $2 \times (-25.1 \text{ kJ/mol})$ or -50.2 kJ/mol. The total energy released is -133 kJ/mol + (-50.2 kJ/mol) = -183.2 or -183 kJ/mol.

$$\text{The percent efficiency of energy capture} = \frac{2(-30. \text{ kJ/mol}) \times 100}{-183 \text{ kJ/mol}} = 33\%$$

This indicates that the efficiency is less than that of glycolytic catabolism of glucose that does not include the conversion of pyruvate to lactate.

4. The large negative standard free energy change for pyruvate formation (step 10), -31.4 kJ/mol, helps drive glycolysis as a whole. The glycolytic process has a $\Delta G^{\circ\prime}$ of -72.8 kJ/mol as determined in problem D.1. Under physiological conditions, the change in free energy, ΔG, is even more negative than is $\Delta G^{\circ\prime}$.

As product is formed in the aldolase reaction (step 4), it is used in the next pathway step. This prevents a disadvantageous equilibrium from being established at the aldolase step.

5. Enolase catalyzes the conversion of 2-phosphoglycerate to phosphoenolpyruvate (step 9). With enolase deficiency, 2-phosphoglycerate accumulates, as do glycolytic intermediates of previous steps, including 1,3-bisphosphoglycerate. 2,3-Bisphosphoglycerate modulates the affinity of hemoglobin for oxygen. The greater the concentration of 2,3-bisphosphoglycerate, the lower the affinity of hemoglobin for oxygen, which results in less oxygen delivered to the tissues. Text Figure 11.24 illustrates the formation and degradation of 2,3-bisphosphoglycerate, which involves the nonglycolytic enzymes bisphosphoglycerate

mutase and 2,3-*bis*phosphoglycerate phosphatase, respectively. A buildup of 1,3-*bis*phosphoglycerate results in the formation of more 2,3-*bis*phosphoglycerate.

Chapter 12

A. True-False

1. <u>True</u>.   When this occurs, the overall stoichiometry involves the input of 6 molecules of glucose 6-phosphate and the production of 12 NADPH, 6 CO_2, and the regeneration of 5 molecules of glucose 6-phosphate.

2. <u>False</u>.   High activity (especially in rapidly dividing cells) of the pentose phosphate pathway is required to supply ribose 5-phosphate for nucleotide biosynthesis.

3. <u>True</u>.   It is formed by the catalytic action of glycogen phosphorylase on glycogen.

4. <u>False</u>.   Glycogen synthase does not catalyze the formation of the α-(1→6) linkages required for formation of branches. This is accomplished by the enzyme amylo-(1,4→1,6)-transglycosylase (the branching enzyme). In addition, glycogenin catalyzes primer extension.

5. <u>False</u>.   Glycogen synthase requires an existing polymer of four to eight glucose residues as a primer. See text Section 12.5 A.

6. <u>False</u>.   While some reactions are reversals of glycolytic reactions, four different enzymes are required to bypass the three metabolically irreversible reactions of glycolysis.

7. <u>False</u>.   Four more moles of ATP are required for gluconeogenesis from two moles of pyruvate than are formed by glycolysis.

8. <u>True</u>.   It involves glycolysis in peripheral tissues and gluconeogenesis in the liver.

9. <u>True</u>.

10. <u>False</u>.   It is produced in the oxidative stage.

11. <u>True</u>.   Hence, normal blood glucose levels are maintained at all times, even at the expense of peripheral (muscle) tissue. (Note: The brain does utilize ketone bodies during starvation.)

12. <u>True</u>.   The glucose is delivered to brain cells, adipocytes, and erythrocytes.

13. <u>True</u>.   Only liver cells are rich in glucagon receptors.

B. Short Answers

1. NADPH

2. transaldolase

3. gluconeogenesis; pentose phosphate pathway; gluconeogenesis;glycogen synthesis

4. limit dextrin

5. pyruvate carboxylase; phosphoenolpyruvate carboxykinase; fructose 1,6-bisphosphatase; glucose 6-phosphatase

6. glycogenin

7. insulin; glucagon; epinephrine

8. Hormonal induction

9. lactate; alanine (and most other amino acids); glycerol

10. substrate cycle

11. liver; kidney

C. Problems

1. Equation 12.1 in your text shows that 4 ATP, 2 GTP, and 2 NADH are *consumed* as one glucose is produced from two pyruvate molecules.

2. Glycolysis produces a net 2 ATP and 2 NADH. Comparing this with the gluconeogenesis energy consumption indicated in the previous answer, 4 ATP equivalents are consumed by the two processes.

3. Yes, the CO_2 carried by the biotin prosthetic group of pyruvate carboxylase attaches to C-3 of pyruvate, thus becoming C-4 of the product, oxaloacetate. In the PEPCK-catalyzed reaction, C-4 of oxaloacetate is released as CO_2 as phosphoenolpyruvate is formed.

4. Because the liver and intestine are perfused in series, food fuels absorbed by the intestine flow immediately through the liver, which regulates the distribution of dietary fuels to other tissues.

5. In Figure 12.1, the sigmoidal shape of the curve in the absence of AMP indicates positive cooperativity of binding of the substrate P_i. AMP enhances the reaction rate by lowering the apparent K_m for P_i. Thus, glycogen phosphorylase appears to be allosterically regulated, with AMP serving as a positive modulator. (Note: Glycogen phosphorylase exists in two active forms, glycogen phosphorylase *a*, a phosphorylated dimer, and glycogen phosphorylase *b*, a less active, nonphosphorylated dimer. These are the R (active) forms of the enzyme, which can also exist in T (inactive) forms. Glycogen phosphorylase *b* is formed when AMP binds to the nonphosphorylated T form of the enzyme.)

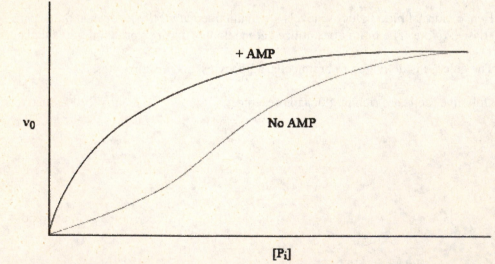

Figure 12.1 Effect of AMP on Glycogen Phosphorylase, with Constant Glycogen Concentration

6. The reactions bypassed are the three metabolically irreversible reactions of glycolysis. Since they are not reversible, they must be bypassed by these four reactions specific to gluconeogenesis.

7. Glycogenolysis results in the production of glucose 1-phosphate and thence glucose 6-phosphate in both tissues. In muscle, the glucose 6-phosphate is metabolized for energy production via glycolysis and the citric acid cycle. In liver, most of the glucose 6-phosphate is converted to glucose, which leaves the cells and is delivered to other tissues, such as the brain, for their energy needs.

8. Glycogen synthase catalyzes the addition of glucose residues from UDP-glucose to the nonreducing end of a glycogen chain. The branching enzyme, amylo-$(1,4 \rightarrow 1,6)$-transglycosylase, catalyzes the removal of an oligosaccharide segment of at least six residues from the nonreducing end of a glycogen chain and attachment of this segment by an α-$(1 \rightarrow 6)$ linkage to the chain. This action creates a branch on the glycogen chain that can be extended by the action of glycogen synthase.

9.

oxaloacetate   phosphoenolpyruvate (PEP)

(b) Prolonged release of glucagon causes continued elevation of cAMP, which triggers increased transcription of the PEP carboxykinase gene in liver, resulting in increased synthesis of PEP carboxykinase. The rate of gluconeogenesis is increased as a consequence of increased enzyme concentration.

10. (a) The Cori cycle is a combination of glycolysis that occurs in muscle and other peripheral tissues and gluconeogenesis that occurs in liver. Lactate produced by glycolysis in peripheral tissues is carried to the liver via the blood where it is converted to glucose by gluconeogenesis.

(b) The function of the cycle is to allow continued energy production in peripheral tissues via glycolysis. The glucose produced in the liver by gluconeogenesis is transported back to the peripheral tissues via the blood for continued glycolytic metabolism. The Cori cycle helps maintain the level of glucose in the blood.

11. (a) The pentose phosphate pathway is most active in mammary glands, liver, adrenal glands, and adipose tissue.

(b) The enzymes of the pentose phosphate pathway are located in the cytosol.

(c) The principal functions of the pathway are to produce NADPH, required for reductive biosynthesis, and ribose 5-phosphate, which is required for biosynthesis of nucleotides (and, hence DNA and RNA) and related derivatives.

12. (a) 6-Phosphogluconate dehydrogenase catalyzes the oxidative decarboxylation of 6-phosphogluconate to form CO_2, NADPH, and ribulose 5-phosphate.

(b) This reaction occurs in the pentose phosphate pathway and is the last step in the oxidative stage of the pathway.

13. (a) xylulose 5-phosphate + ribose 5-phosphate

(b) xylulose 5-phosphate + erythrose 4-phosphate

(c) sedoheptulose 7-phosphate + glyceraldehyde 3-phosphate

(Note that these are reversals of reactions of the pentose phosphate pathway. Such reversals are possible since all reactions in the nonoxidative stage of the pathway are near equilibrium reactions.)

14. A supply of reducing power (NADPH) from the pentose phosphate pathway is necessary for maintaining iron in its reduced form, Fe^{2+}, for hemoglobin synthesis.

D. Additional Problems

1. (a) From the data in study guide Table 11.1, add the standard free energy changes, $\Delta G^{\circ\prime}$, for the **reverse** of glycolysis reactions 2, 4, 5, 6, 7, 8 and 9 as well as the standard free energy changes for the four reactions unique to gluconeogenesis. Remember to multiply by two the $\Delta G^{\circ\prime}$ values of those reactions in the triose stage and to change the signs of the $\Delta G^{\circ\prime}$ values in study guide Table 11.1 for the reversed glycolysis reactions.

$\Delta G^{\circ\prime} = [-1.67 + (-24.0) + (-7.66) + (2 \times -6.28) + (2 \times 18.8) + (2 \times -4.44) + (2 \times -1.84) + (2 \times 2.1) + (-16.7) + (-13.8)]$ kJ/mol
$= [-88.95 + 41.8]$ kJ/mol
$= -47.2$ kJ/mol.

(b) The process is spontaneous as indicated by the negative value of the standard free energy change.

2. The process would not be spontaneous. You determined in study guide problem 11.D.1. that the $\Delta G^{\circ\prime}$ for glycolysis was -72.8 kJ/mol. If all glycolysis reactions were reversed, the $\Delta G^{\circ\prime}$ for the process would be +72.8 kJ/mol. This process, with its large positive value of $\Delta G^{\circ\prime}$, would not be spontaneous. Bypassing the three metabolically irreversible reactions of glycolysis with reactions that, overall, have a reasonably large negative change in standard free energy allows gluconeogenesis to be spontaneous.

3. (a)

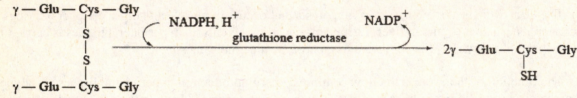

(b) Glucose 6-phosphate dehydrogenase is the enzyme that catalyzes the first reaction of the pentose phosphate pathway.

(c) Glutathione is required for proper maintenance of red blood cell membranes, as described in the question. Replenishing glutathione from the dimer form requires NADPH, which is produced by the pentose phosphate pathway. A deficiency of glucose 6-phosphate dehydrogenase could lead to insufficient NADPH production that would also limit the amount of glutathione available for red blood cell maintenance.

4. One effect of caffeine is the inhibition of cAMP phosphodiesterase, which prolongs the epinephrine-producing effect of cAMP. Epinephrine (through an increase of cAMP) leads to glycogen degradation to provide glucose for energy. Glycogen phosphorylase *a* is the active form of the enzyme required for glycogen degradation as indicated in text Figure 12.16. Inhibition of this enzyme by caffeine should prevent glycogen degradation. It seems, therefore, that the two actions are opposed to each other.

5. (a) In a unit of the glycogen chain that includes a branch, there would be 16 residues (8 along the main chain and 8 in the branch). Dividing the total residues (6000) by 16 per unit gives 375. So, there are about 375 ends.

(b) Glycogen molecules are synthesized with the reducing end of a short segment attached to the protein glycogenin. Extension of chains and branches is toward the nonreducing ends, since carbon 1 of the residue being attached is connected to carbon 4 of the residue of the existing chain. Therefore, all ends of the glycogen molecule are nonreducing ends.

6. Glucose is reduced to sorbitol by NADPH and aldose reductase. Sorbitol is then oxidized to fructose by NAD^+ and polyol dehydrogenase.

Chapter 13

A. True-False

1. <u>True.</u>   FAD is a prosthetic group of an enzyme of the succinate dehydrogenase complex that accepts hydrogens from succinate and immediately transfers them to coenzyme Q, the mobile carrier of reducing power within the inner mitochondrial membrane.

2. <u>True.</u>   The citric acid cycle generates six molecules of CO_2 for each molecule of glucose catabolized.

3. <u>False.</u>   Your text points out that the cycle is a multistep catalyst. Since the process is cyclic, the intermediates are continually recycled. However, sufficient amounts of oxaloacetate must be present to allow appropriate rates of acetyl CoA uptake.

4. <u>False.</u>   There are two six-carbon intermediates, one five-carbon intermediate, and five four-carbon intermediates as indicated in Figure 13.1. To better familiarize yourself with these intermediates, make a list of them by name.

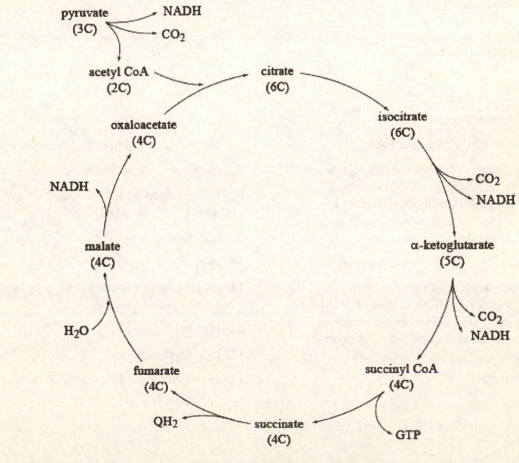

Figure 13.1 The Citric Acid Cycle

5. <u>False.</u>   They are located in mitochondria.

6. <u>False</u>.   Pyruvate does pass freely through porins in the outer mitochondrial membrane, but its passage, in symport with H$^+$, through the inner mitochondrial membrane is aided by pyruvate translocase.

7. <u>True</u>.   It is several times larger than a ribosome.

8. <u>True</u>.

9. <u>True</u>.   Succinyl CoA condenses with glycine to initiate porphyrin biosynthesis, but succinyl CoA can be replenished by degradation of some amino acids.

10. <u>False</u>.   Although coenzyme A is required in two different steps as pyruvate is catabolized, two moles of coenzyme A are formed as products in the citric acid cycle, so there is no net consumption of coenzyme A..

11. <u>False</u>.   In addition to the availability of acetyl CoA and oxaloacetate to initiate citrate formation, control is exercised through allosteric modulators and their effects on the cycle enzymes citrate synthase, isocitrate dehydrogenase, and the α-ketoglutarate dehydrogenase complex.

12. <u>True</u>.   Glycolysis produces two ATP per mole of glucose. From the two moles of pyruvate produced by glycolysis, the citric acid cycle produces two moles of GTP (or ATP).

13. <u>True</u>.   The prochiral molecule contains a carbon of the type C_{aacd}, which can be converted to a chiral molecule by replacement of one of the *a* groups by a group different from the remaining three groups. The carbon would then be C_{abcd}.

14. <u>True</u>.   Both are complexes of several different enzymes and utilize the same cofactors (TPP, lipoamide, FAD, CoASH, and NAD$^+$).

15. <u>False</u>.   It is the step catalyzed by succinyl CoA synthase that results in substrate level phosphorylation as GTP (or ATP) is formed.

B. Short Answers

1. coenzyme A; NAD$^+$; lipoate, FAD; TPP

2. succinate; fumarate; malate; oxaloacetate (Prior to the elucidation of the citric acid cycle, biochemist Albert Szent-Gyorgyi discovered that these dicarboxylic acids stimulate respiration.)

3. CO_2; NAD$^+$

4. the pyruvate dehydrogenase complex [This complex contains multiple copies of pyruvate dehydrogenase (E1), dihydrolipoamide acetyltransferase (E2), and dihydrolipoamide dehydrogenase (E3).]

5. citrate synthase

6. the α-ketoglutarate dehydrogenase complex

7. succinyl CoA; succinate

8. five (including the conversion of pyruvate to acetyl CoA); NADH; QH2

9. malonate

10. two

11. α-ketoglutarate; succinyl CoA (as well as citrate and oxaloacetate)

12. NADH

13. L-malate; fumarate

14. α-ketoglutarate; succinyl CoA

C. Problems

1. Coenzyme A is composed of an ADP moiety, an esterified phosphate, a pantothenic acid unit, and a mercaptoethylamine unit. Coenzyme A functions as a carrier of acyl groups such as the acetyl group formed when pyruvate is oxidatively decarboxylated by the pyruvate dehydrogenase complex, and the

succinyl group formed when α-ketoglutarate is oxidatively decarboxylated by the α-ketoglutarate dehydrogenase complex.

2. Sources of acetyl CoA are pyruvate (from glucose and other sources), fatty acids (from triacylglycerols and other lipids), and the carbon skeletons of some amino acids.

3. Pyruvate, produced in the cytosol by glycolysis, enters mitochondria through porins in the outer membrane and by the aid of pyruvate translocase in the inner membrane. In the mitochondrial matrix, pyruvate is converted to acetyl CoA, a substrate for the citric acid cycle.

4. The products are released sequentially. In the five-step process, pyruvate is first decarboxylated to form CO_2. Acetyl CoA is formed as the second product and NADH is formed last as the enzyme complex is regenerated.

5. Citrate is a prochiral molecule. It is bound to aconitase such that stereospecific addition of water to the C=C of the *cis*-aconitate intermediate results in the formation of isocitrate (with two chiral centers).

6. A kinase and a phosphatase are part of the mammalian pyruvate dehydrogenase complex. Pyruvate dehydrogenase kinase catalyzes the phosphorylation of E1, pyruvate dehydrogenase, causing its inactivation. Pyruvate dehydrogenase phosphatase catalyzes the dephosphorylation and activation of E1. Control of the complex is achieved by control of E1, since it catalyzes the rate-determining step in the conversion of pyruvate to acetyl CoA and CO_2.

7. (a) An abundance of ADP stimulates isocitrate dehydrogenase so that more ATP is ultimately formed from the reduced cofactors NADH and QH_2 produced by the cycle. ADP also inhibits pyruvate dehydrogenase kinase, allowing pyruvate dehydrogenase to remain active.

    (b) Available acetyl CoA cannot all be converted to citrate when oxaloacetate levels are too low. More oxaloacetate is formed by the carboxylation of available pyruvate.

8. (a) four NADH, one QH_2, one GTP

    (b) three NADH, one QH_2, one GTP

    (c) eight NADH, two QH_2, two GTP (twice that of pyruvate since two moles of pyruvate are produced from each mole of glucose)

9.

$$
\begin{array}{c}
CH_2-COO^- \\
|\\
CH-COO^- \\
|\\
HO-CH-COO^-
\end{array}
+ 2\,NAD^+ + GDP + P_i \longrightarrow 2\,CO_2 + 2\,NADH + H^+ + GDP +
\begin{array}{c}
CH_2-COO^- \\
|\\
CH_2-COO^-
\end{array}
$$

10. Pigeon muscle tissue contains citric acid cycle enzymes and cofactors but malonate inhibits the succinate dehydrogenase complex, causing succinate to accumulate.

11. Reactions *in vivo* are not at equilibrium, but are, instead, in a steady state of formation and consumption of intermediates. As isocitrate is formed in the cell, it is used in the next step of the citric acid cycle and more isocitrate is rapidly formed from citrate.

12. Accumulation of acetyl CoA partially reverses the reactions catalyzed by E2 and E3. Acetyl CoA also activates pyruvate dehydrogenase kinase in mammals. If sufficient product (acetyl CoA) of the step catalyzed by the complex is available, the enzyme is modulated to decrease pathway flux.

13. Succinate + Q + H_2O + NAD^+ $\rightarrow$ oxaloacetate + QH_2 + NADH + H^+

D. Additional Problems

1. Due to the pyruvate dehydrogenase deficiency, pyruvate and, consequently, lactate accumulate in the patient's serum, causing acidosis. Although less than normal amounts of acetyl CoA are produced from pyruvate, fatty acid oxidation yields acetyl CoA, as do some of the amino acids. Acetyl CoA from these sources fuels the citric acid cycle. Lowering carbohydrate levels in the diet would help minimize the pyruvate and lactate accumulation in the blood.

2. The labeled acetyl CoA from fatty acid catabolism condenses with oxaloacetate to form citrate. Two molecules of CO_2 are subsequently lost from citrate in the citric acid cycle, but the carboxyl groups lost as CO_2 are those from the original oxaloacetate and not from the acetyl CoA. Therefore, the labeled carbons are incorporated in the four-carbon intermediates including succinate and, ultimately, oxaloacetate. Since oxaloacetate is a precursor for glucose via gluconeogenesis, some of the labeled carbons can be found in glucose. There is no net synthesis of glucose since two carbons, as acetyl CoA, go into the cycle and two carbons, as CO_2, are produced.

3.

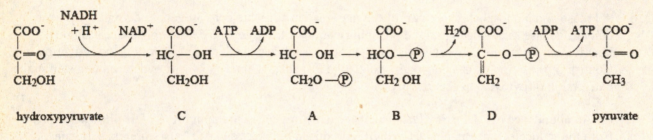

4. Pyruvate carboxylase catalyzes the carboxylation of pyruvate to form oxaloacetate, an anaplerotic reaction that replenishes oxaloacetate for use in the citric acid cycle and gluconeogenesis. In this diseased state, oxaloacetate cannot be replenished at normal levels. This results in lower than normal ATP production, due to diminished activity of the citric acid cycle, and lower than normal glycogen stores available for energy production, due to diminished gluconeogenesis. Therefore, the afflicted individual has limited capacity for prolonged exercise.

5. (a) $\Delta G^{\circ\prime}$ = $-RT\ln K_{eq}$
   = $-(8.315 \times 10^{-3}$ kJ mol^{-1} K$^{-1})(298$ K$)(\ln 5 \times 10^5)$
   = -32.5 kJ mol^{-1}.

   (b) The second reaction is essentially a reversal of the first, so the $\Delta G^{\circ\prime}$ for the formation of oxaloacetate and acetyl CoA from citrate and CoASH would be $+32.5$ kJ mol^{-1}. However, the hydrolysis of ATP to ADP and P_i is also involved. Text table 10.1 gives $\Delta G^{\circ\prime} = -30$ kJ mol^{-1} for this reaction. Combination of these two values gives $\Delta G^{\circ} = +2.5$ kJ mol^{-1} for the overall reaction. Since $\Delta G^{\circ\prime} = -RT\ln K_{eq}$,

   $\ln K_{eq}$ = $\Delta G^{\circ\prime}/-RT$
   = $+2.5$ kJ mol$^{-1}/(-8.315 \times 10^{-3}$ kJ mol^{-1} K$^{-1})(298$ K$)$
   = -1.0
   $K_{eq} = 0.37$

   (c) The large equilibrium constant for the citrate formation reaction and its negative $\Delta G^{\circ\prime}$ suggest that the reverse process would not proceed spontaneously. In the reaction catalyzed by citrate lyase, energy is provided by the net hydrolysis of one high-energy bond of ATP. The ATP is required for synthesis of the thiol ester intermediate, citryl CoA. When citrate accumulates and some is transported to the cytosol, sufficient ATP is available and not as much citrate must be catabolized by the citric acid cycle. Some ATP, therefore, can be expended to lyse citrate, forming acetyl CoA, which can be used for synthesis of fatty acids that can be stored for future energy needs.

Chapter 14

A. True-False

1. <u>False</u>.   The outer membrane has few proteins, but the inner membrane is rich in proteins.

2. <u>True</u>.

3. <u>True</u>.   Since oxidative phosphorylation is coupled to electron transport, both occur under normal conditions. Electron transport drives ATP formation.

4. <u>True</u>.   Molecular O_2 is reduced to 2 H_2O and 4 protons are translocated across the inner mitochondrial membrane by Complex IV.

5. <u>False</u>.   Q (ubiquinol) is a lipid.

6. <u>False</u>.   Iron sulfur clusters are found only in Complexes I, II, and III.

7. <u>False</u>.   Proton concentration increases in the mitochondrial intermembrane space.

8. <u>False</u>.   Since the two processes are coupled, the inhibition of one stops the other.

9. <u>True</u>.   See section 14.8 of the text.

10. <u>True</u>.   2,4-Dinitrophenol is an uncoupler that disconnects ATP production from electron transport, thus allowing faster oxygen uptake.

11. <u>False</u>.   Oxidation of 1 mole of NADH produces about 2.5 moles of ATP, but 1 mole of succinate produces only about 1.5 moles of ATP.

12. <u>True</u>.   According to the chemiosmotic theory, a proton concentration gradient is created by the transfer of protons from the matrix to the intermembrane space of the mitochondrion as electrons are transferred by the electron transport system.

13. <u>True</u>.   In mitochondria, the reverse reaction (formation of ATP from ADP and Pi) occurs, but the F_1-ATPase activity was first detected in knobs isolated from mitochondria.

14. <u>False</u>.   Complex II does not contribute to the proton concentration gradient.

15. <u>False</u>.   Three of the four H^+ pass through the channel, but the fourth H^+ is transported in a symport process with phosphate ($H_2PO_4^-$).

B. Short Answers

1. cytochromes

2. electron-transporting complexes (respiratory complexes)

3. four; cofactors

4. Q; cytochrome c

5. Q (ubiquinone)

6. cyanide; carbon monoxide (other inhibitors are also known)

7. Complex II; succinate dehydrogenase

8. stalk and knob structures of the inner mitochondrial membrane (Complex V, ATP synthase)

9. Peter Mitchell; chemiosmotic theory

10. P:O ratio

11. adenine nucleotide translocase

C. Problems

1. Four protons are translocated to the intermembrane space per pair of electrons that pass from NADH to Q, forming QH_2.

2. $2\ NADH + 7\ H^+ + 5\ ADP + 5\ P_i + O_2 \rightarrow 2\ NAD^+ + 5\ ATP + 7\ H_2O$

   (Note: This is based on the ionic states and conversions shown in text Figure 10.9.)

3. $2\ succinate + 3\ ADP + 3\ P_i + 3\ H^+ + O_2 \rightarrow 3\ ATP + 2\ fumarate + 5\ H_2O$

   (Note: This is based on the ionic states and conversions shown in text Figure 10.9.)

4. Succinate transfers its electrons to Q (ubiquinone) with no contribution to the formation of the proton concentration gradient. Because the electrons from succinate enter the electron transport chain at a point that bypasses Complex I, the free energy that becomes available when a substrate enters at Complex I is not obtained. Only the free energy obtained from Complexes III and IV contributes to the proton concentration gradient that drives the formation of ATP. (This is consistent with the requirement of a stronger oxidizing agent than NAD^+ to oxidize.)

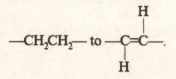

5. Use the standard reduction potentials from text Table 10.4 as follows:

   $$NAD^+ + 2\ H^+ + 2\ e^- \rightarrow NADH + H^+ \qquad E^{\circ\prime} = -0.32\ V$$
   $$FMN + 2\ H^+ + 2\ e^- \rightarrow FMNH_2 \qquad E^{\circ\prime} = -0.22\ V$$

   The equation for the oxidation of NADH and the cell potential is:

   $$NADH + H^+ + FMN \rightarrow NAD^+ + FMNH_2 \qquad \Delta E^{\circ\prime} = +0.10\ V$$

   From this, the standard free energy change can be calculated from:

   $$\Delta G^{\circ\prime} = -nF\Delta E^{\circ\prime}, \text{ where } n = 2;\ F = 96.48\ kJ\ V^{-1}\ mol^{-1};$$
   $$\Delta G^{\circ\prime} = -(2)(96.48\ kJ\ V^{-1}\ mol^{-1})(0.10\ V)$$
   $$= -19.3\ kJ/mol = -19\ kJ/mol.$$

6. A respiratory chain inhibitor stops both electron transfer and oxidative phosphorylation, since there is no production of a proton concentration gradient. The inhibition site of antimycin A is between cytochrome b_{560} and $\cdot Q^-$ or QH_2. Therefore, (a) oxygen is not consumed, (b) no ATP is formed, (c) neither NAD^+ nor Q is regenerated, and (d) there is no heat production.

7. 2,4-Dinitrophenol uncouples the process of electron transport from the synthesis of ATP. Therefore, (a) oxygen consumption is increased since electron transfer can occur more rapidly, (b) no ATP is formed, (c) NAD^+ is regenerated, and (d) heat production is increased since the energy normally used for ATP formation is released as heat.

8. Text Table 14.2 gives the standard free energy changes, with NADH as the substrate, for electron transfer involving each of the four complexes. The standard free energy change for the entire process is:

   $$\Delta G^{\circ\prime}_{Total} = [(-70) + (-37) + (-110)]\ kJ/mol$$
   $$= -217\ kJ/mol.$$

This is essentially the same value determined in text Section 10.9.B. Note that the standard free energy change shown for Complex II (-2 kJ/mol) was not included in this tabulation since Complex II does not accept electrons from NADH.

9. $(4H^+)(-19.4 kJ/H^+) = -77.6$ kJ; $(-77.6 kJ)(1$ ATP$/-30 kJ) = 2.59$ ATP possible;
    percent captured per 4 $H^+ = (1$ ATP$/2.59$ ATP$) \times 100 = 38.6\%$

10. 2,4-Dinitrophenol dissipates the proton concentration gradient. It can release its proton in the basic mitochondrial matrix. The anionic form is sufficiently hydrophobic that it can diffuse through the inner membrane to the cytosol, pick up another proton, move back to release the proton into the matrix, and repeat the process. Thus, the matrix becomes less basic than in the absence of 2,4-dinitrophenol and the proton concentration gradient between the matrix and the intermembrane space is diminished or lost.

11. A continuing supply of ADP and P_i from the cytosol is required to make ATP. Synthesized ATP must be transported from mitochondria for use in other areas of the cell. Carrier proteins in the inner mitochondrial membrane transport the nucleotides into and out of the mitochondria.

12. Oligomycin binds in the channel of the F_0 component of Complex V, preventing the entry of protons necessary for ATP synthesis.

13. ATP cannot diffuse from the mitochondrial matrix to the cytosol. A transporter called adenine nucleotide translocase exchanges mitochondrial ATP for cytosolic ADP. P_i is also transported into mitochondria. The combined transport process requires energy equivalent to one proton of those translocated.

D. Additional Problems

1. (a) Because 2,4-dinitrophenol is an uncoupling agent, electron transport continues but little or no ATP is formed. Lowered ATP levels signal the body to mobilize fat stores for energy (ATP) production. (b) Weakness, collapse, and death may occur, depending on the severity of ATP depletion. Oxygen consumption is increased as the burden of ATP formation is removed by uncoupling of the electron transport system. Energy normally used to drive ATP formation is released as heat, elevating body temperature and resulting in profuse sweating in response to this condition.

2. (a) $O_2 + 4 e^- + 4 H^+ \rightarrow 2 H_2O$

    (b) As indicated in text Figure 14.13, two protons are translocated from the mitochondrial matrix to the intermembrane space for each pair of electrons transferred, which results in the reduction of one atom of oxygen to water. The formation of water, the product of oxygen reduction, requires two protons, which are used from those in the mitochondrial matrix. The net result is that the concentration of protons in the intermembrane space is four times greater than that in the mitochondrial matrix.

3. You determined in problem C.5. that the change in standard free energy for the transfer of electrons from NADH to the FMN of Complex I is -19 kJ/mol. Formation of one ATP requires 30 kJ/mol. However, for transfer of electrons to Q,

    $$NADH + H^+ + Q \rightarrow QH_2 + NAD^+, \qquad E^{o'} = +0.32 \text{ V} + 0.04 \text{ V} = +0.36 \text{ V}$$
    $$\Delta G^{o'} = -nF\Delta E^{o'} = -(2)(96.48 \text{ kJ V}^{-1} \text{ mol}^{-1})(0.36 \text{ V})$$
    $$= -69.47 = -69 \text{ kJ/mol}.$$

    This is more than enough energy for the formation of one ATP.

4. Thermogenesis is the term for heat generation. Brown adipose tissue (brown fat) is rich in mitochondria. In these mitochondria, long-chain fatty acids, released in response to norepinephrine, uncouple oxidative phosphorylation from electron transport. The energy thus available is released as heat to maintain body temperature.

5. Since neither NADH nor succinate is oxidized, neither Complex I nor Complex II is present. Complex IV catalyzes the transfer of electrons from reduced cytochrome *c* to oxygen. Therefore, Complex IV is present, but there is no evidence for the presence or absence of Complex III.

Chapter 15

A. True-False

1. <u>True</u>. These organisms are dependent, in the dark, on mitochondrial respiration for energy.

2. <u>True</u>. Although thermodynamically unfavorable, the oxidation of water is driven by input of solar energy.

3. <u>False</u>. Although the dark reactions do not depend on light to occur, both the light and dark reactions of photosynthesis occur simultaneously.

4. <u>False</u>. Oxygen is produced by the light-dependent reactions.

5. <u>False</u>. Labeling experiments have shown that CO_2 taken in by the plant is fixed as carbohydrate and that oxygen is produced from the oxidation of water. (See Section 15.4 and Figure 15.18 in your text.)

6. <u>True</u>. The accessory pigments present in photosynthetic membranes complement the absorptions of chlorophyll *a* and chlorophyll *b* so that light across the visible spectrum is absorbed and used.

7. <u>False</u>. Green and purple bacteria use bacteriochlorophyll *a* and bacteriochlorophyll *b* for this function.

8. <u>False</u>. Protons are pumped into the lumen, the aqueous space enclosed by the thylakoid membrane.

9. <u>True</u>. Plants contain two units, photosystem I and photosystem II, which have slightly different absorption maxima and electron-transport chains.

10. <u>True</u>. For this reason, the cyclic electron transport sequence operates to form ATP without the simultaneous formation of NADPH. (See text Section 15.4.)

11. <u>False</u>. Sucrose is formed in the cytosol, but starch is formed in chloroplasts and amyloplasts.

B. Short Answers

1. light; O_2; ATP; NADPH; dark; carbohydrates
2. chloroplasts
3. thylakoid; grana
4. light-harvesting complexes
5. photophosphorylation
6. reaction center; special pair
7. thylakoid membrane
8. P680$^+$ (special pair chlorophylls that absorb @ 680 nm)
9. ferredoxin
10. oxygen-evolving complex
11. stroma
12. 3-phosphoglycerate
13. NADPH
14. ribulose-1,5-*bis*phosphate carboxylase; Rubisco
15. PSI; PSII; cytochrome *bf*; ATP synthase

C. Problems

1. Due to its hydrophobic nature, the phytol side chain helps anchor chlorophyll molecules in membranes.

2. Chlorophyll exists in the ground state, Chl^o; in an excited state, Chl^*, after absorption of a photon of light; and in the oxidized state, Chl^+.

3. In the light reactions, solar energy is used to oxidize water to produce oxygen. Protons derived from water are used in the chemiosmotic synthesis of ATP from ADP and P_i, and electrons from water reduce $NADP^+$ to NADPH. In the dark reactions, ATP and NADPH provide the energy and reducing power, respectively, to convert CO_2 to carbohydrates.

4. Sucrose is made from four moles of triose phosphate (shown below as G 3-P), each of which requires three moles of CO_2. Two NADPH are required to reduce each mole of CO_2, and two electrons are transferred from each mole of water to reduce $NADP^+$ to NADPH. Therefore,

$$\text{moles } H_2O = (H_2O/2 \text{ e}^-)(2 \text{ e}^-/NADPH)(2NADPH/CO_2)(3 \text{ } CO_2/G \text{ 3-P})(4 \text{ G 3-P/sucrose})$$
$$= 24 \text{ moles } H_2O/\text{mole sucrose}.$$

5. Since $E = h\nu$, and $\nu = c/\lambda$, then $E = h \text{ } c/\lambda$. The energy of a mole of photons of 400 nm light is:

$$E = [6.626 \times 10^{-34} \text{ J s}][(3.00 \times 10^{17} \text{ nm/s})/400 \text{ nm}]$$
$$= 4.97 \times 10^{-19} \text{ J/photon} \times 6.02 \times 10^{23} \text{ photons/mol}$$
$$= 2.99 \times 10^5 \text{ J/mol} = 299 \text{ kJ/mol}.$$

The energy of a mole of photons of 700 nm light is:

$$E = [6.626 \times 10^{-34} \text{ J s}][(3.00 \times 10^{17} \text{ nm/s})/700 \text{ nm}]$$
$$= 2.840 \times 10^{-19} \text{ J/photon} \times 6.02 \times 10^{23} \text{ photons/mol}$$
$$= 1.71 \times 10^5 \text{ J/mol} = 171 \text{ kJ/mol}.$$

The photon of light of shorter wavelength, 400 nm, does provide a greater amount of energy.

6. The carotenoids are accessory pigments or antenna pigments that, like chlorophyll, are capable of absorbing light. They absorb light of wavelengths not absorbed maximally by chlorophyll and transfer the absorbed energy to adjacent pigments in the photosystem. The energy is ultimately transferred to the special pair chlorophylls in the reaction center.

7. Similarities include:

   a. location within a cellular organelle

   b. membrane-bound components

   c. transfer of electrons via system components

   d. production of a proton concentration gradient

   e. formation of ATP

   f. an ATP synthase with a stalk and knob structure

8. If 8 photons produce 1 O_2, and 1 photon excites only 1 in 2400 chlorophyll molecules, (8 photons/O_2) × (O_2/2400 chlorophyll) = 1 photon per 300 chlorophyll. This indicates that 300 chlorophyll molecules constitute a functional cluster, which utilizes one photon. It must also indicate that eight such clusters are necessary to produce one O_2 molecule.

9. It takes 2 molecules of water to produce 1 oxygen molecule, 2 ATP, and 2 NADPH molecules. Tripling this process would provide 12 ATP and 12 NADPH. Fixation of 6 CO_2 as 1 hexose requires 18 ATP and 12 NADPH. Thus, it appears there is a deficiency of 6 ATP. More ATP is formed without concomitant NADPH formation by cyclic electron transport, as described in text Section 15.4. High NADPH concentrations favor cyclic electron flow.

10. PSI is located in the stroma thylakoid and PSII is located in the grana thylakoid membrane. These differences in location prevent direct transfer of excitation energy between the two photosystems, which would short circuit energy transfer. They are linked by specific electron carriers and work in series.

11. This study supported Mitchell's theory regarding ATP formation via chemiosmosis. The artificially generated proton concentration gradient drives the formation of ATP. It suggested that the same kind of process occurs when ATP is formed during photophosphorylation (i.e., illumination causes accumulation of protons in the lumen).

12. (a) Six molecules of CO_2 are required.

(b) Each of the 6 CO_2 molecules combines with a ribulose-1,5-*bis*phosphate molecule, forming a total of 12 3-phosphoglycerate molecules. Of these 12, 2 are used to produce 1 hexose and 10 go through reactions of the pentose phosphate pathway to regenerate 6 ribulose-1,5-*bis*phosphate molecules.

13. 2-Carboxyarabinitol 1-phosphate binds to the carbamoylated (active) form of Rubisco, rendering it inactive. In some plants, enough 2-carboxyarabinitol 1-phosphate is synthesized to keep Rubisco completely inactive in the dark.

14. These plants open their stomata at night to take in CO_2 with minimal water loss. This CO_2 is fixed by PEP carboxylase to form oxaloacetate. The oxaloacetate is reduced to malate that is stored in a central vacuole of the plant. During the day, malate is released from the vacuole and decarboxylated. The CO_2 thus freed is available for carbon assimilation during the day.

15. Starch is made in chloroplasts. One ATP is consumed for each glucose unit added to a starch chain.

D. Additional Problems

1. The special pair chlorophyll molecules and those in the antenna complex are structurally identical. However, the special pair chlorophyll molecules are in a different physical environment that causes them to act as an electron sink with respect to the chlorophyll molecules of the antenna complex. Figure 15.2 indicates schematically how energy converts ground state antenna complex chlorophylls to their excited states and how this energy is ultimately transferred to reaction center chlorophylls because the excited states of the reaction center chlorophylls is lower than those of the antenna complex chlorophylls.

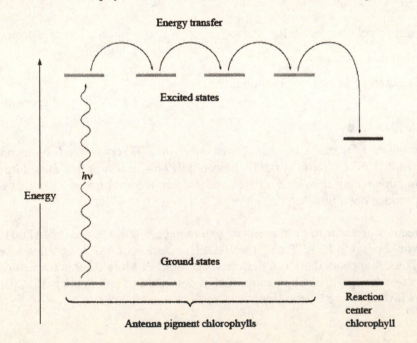

Figure 15.2 Energy Transfer from Antenna Chlorophylls to Reaction Center Chlorophylls

2. Figure 15.1 shows that the maximum yield of ATP occurs at an external pH of about 8.3–8.4. The yield is about 0.2 ATP per chlorophyll or 1 ATP per 5 chlorophyll molecules. Emerson and Arnold found that one photon excited 300 chlorophyll molecules. It takes 8 photons to form 1 O_2 and 2 ATP. Therefore, 8×300 or 2400 chlorophyll molecules produce 2 ATP, which is 1 ATP per 1200 chlorophyll molecules. The much greater ATP yield in the Jagendorf and Uribe study is due to the energy buildup (from the formation of a proton concentration gradient) during the period in which the chloroplasts were exposed to light in the absence of ADP.

3. (a) Your text indicates that, normally, the rate of carboxylation in healthy plants is three to four times that of the oxygenation reaction (photorespiration).

(b) In some plants, photosynthesis, on a hot bright day, will deplete the level of CO_2 at the chloroplast and raise the level of O_2 such that the rate of photorespiration may approach that of photosynthesis. This is a major limiting factor in the growth of many plants. (Note: The oxygenase activity of Rubisco increases with temperature more rapidly than does the carboxylase activity.)

(c) In the C_4 pathway, C_4 acids are decarboxylated in bundle sheath cells, which results in increased CO_2 concentration inside the cells and creates a high ratio of CO_2 to O_2. This minimizes oxygenase activity of Rubisco (photorespiration) and does not limit growth of the plant due to loss of CO_2 that occurs in plants lacking the C_4 pathway. (See text Figure 15.26.)

4. The energy for one hour is:

$$7 \text{ J (cm}^2 \text{ s)}^{-1} \times 10 \text{ cm}^2 \times 3600 \text{ s hr}^{-1} = 252 \text{ kJ/hr.}$$

The potential for glucose formation is:

$$252 \text{ kJ hr}^{-1} \times 1 \text{ mol/2870 kJ} \times 180. \text{ g/mol} = 15.8 \text{ g.}$$

If only 5.53 grams form, the percent efficiency = 5.53 g $\times$ 100/15.8 g = 35%.

5. $\Delta G^{\circ\prime}$ for ATP + $H_2O \rightarrow$ ADP + P_i is -30. kJ/mol. ΔG for ATP synthesis is equal to $\Delta G^{\circ\prime}$ plus the free energy change due to concentrations other than unity. That is:

$$\Delta G = \Delta G^{\circ\prime} + RT \ln([ATP]/[ADP][P_i])$$

But ΔG for the reaction must be equal to three times the free energy change due to transfer of one proton:

$$\Delta G = 3(2.3 \, RT \, \Delta pH + ZF\Delta\Psi).$$

But $\Delta\Psi$ is zero due to the transfer of Ca^{2+} and Cl^- ions in and out of the thylakoid membrane as protons are pumped through the thylakoid membrane. Therefore, the term $ZF\Delta\Psi$ becomes zero. Then,

$$\Delta G^{\circ\prime} + RT \ln Q = 3(2.3 \, RT \, \Delta pH),$$

where Q is the ratio of concentrations, $[ATP]/[ADP][P_i]$. Rearranging gives:

$$\Delta pH = \frac{\Delta G^{\circ\prime} + RT \ln Q}{3(2.3 \, RT)}$$

$$\Delta pH = \frac{30. \text{ kJ/mol} + (8.315 \text{ J mol}^{-1} \text{ K}^{-1})(298 \text{ K})(\ln 1000)}{6.9(8.315 \text{ J mol}^{-1} \text{ K}^{-1})(298 \text{ K})}$$

$$= \frac{(30. + 17.1) \text{ kJ mol}^{-1}}{17.1 \text{ kJ mol}^{-1}} = 2.75 \text{ or } 2.8 \text{ pH units}$$

This shows that a pH gradient in chloroplasts of about 3 pH units exists across the thylakoid membrane.

6. (a) Since 3-phosphoglycerate is a 3-carbon molecule, the acceptor might be a 2-carbon molecule.

(b) The formation of two molecules of 3-phosphoglycerate for each CO_2 suggested that a 5-carbon molecule might be the acceptor and the resulting 6-carbon intermediate was immediately cleaved to two molecules of 3-phosphoglycerate.

(c) The label is found in the carboxyl group of 3-phosphoglycerate. (See text Figure 15.20.)

Chapter 16

A. True-False

1. <u>False</u>.   Carnitine is involved in the transport of fatty acids from the cytosol into mitochondria.

2. <u>True</u>.   Fatty acids with odd-numbered carbon chains are rare.

3. <u>True</u>.   Each two-carbon unit becomes an acetyl CoA.

4. <u>False</u>.   Eicosanoic acid contains 20 carbons and yields 10 acetyl CoA from β-oxidation. Only 9 NADH and 9 QH_2 are formed, since the ninth round of reactions yields 2 molecules of acetyl CoA as the 4-carbon substrate undergoes thiolysis.

5. <u>True</u>.   One ATP is required for the activation. It is converted to AMP and PP_i. Because PP_i is subsequently hydrolyzed to P_i, the cost of fatty acid activation is two high-energy phosphate bonds.

6. <u>False</u>.   β-Oxidation occurs in mitochondria, but fatty acid synthesis occurs in the cytosol.

7. <u>True</u>.   Synthesis occurs in liver mitochondria.

8. <u>False</u>.   The pentose phosphate pathway furnishes about half of the NADPH required for fatty acid synthesis and the rest is furnished by the citrate transport system. (See the previous edition of your text.)

9. <u>True</u>.   The side chain carbon atoms also are derived from acetyl CoA, which is transported from the mitochondria into the cytosol by the citrate transport system.

10. <u>True</u>.   LDL-derived cholesterol can accumulate on the inner arterial walls.

11. <u>True</u>.   This is true both on a per gram and a per carbon basis. Glucose is more oxidized to begin with than is a saturated fatty acid chain, so glucose cannot generate as much NADH and QH_2, hence less ATP.

B. Short Answers

1. triacylglycerols; glycogen

2. acetyl CoA; NADH; QH_2

3. oxidation; hydration; further oxidation; thiolysis

4. 3-ketoacyl-CoA thiolase

5. β-hydroxybutyrate; acetoacetate; acetone

6. mitochondria; peroxisomes

7. malonyl CoA

8. acetyl-CoA carboxylase

9. desaturases

10. arachidonate

11. leukotrienes

12. Lovastatin

13. geranyl pyrophosphate; farnesyl pyrophosphate; squalene

14. endoplasmic reticulum

C. Problems

1. Typically, 16- and 18-carbon fatty acids are made because the condensing enzyme, α-ketoacylACP synthase, cannot accommodate longer chains in its binding site.

2.
$$\text{Acetyl CoA} + 8 \text{ malonyl CoA} + 16 \text{ NADPH} + 16 \text{ H}^+ \rightarrow$$
$$\text{Stearate} + 8 \text{ CO}_2 + 16 \text{ NADP}^+ + 9 \text{ CoASH} + 7\text{H}_2\text{O}$$

3.
$$\text{Stearoyl CoA} + 8 \text{ Q} + 8 \text{ NAD}^+ + 8 \text{ CoASH} + 8 \text{ H}_2\text{O} \rightarrow$$
$$9 \text{ Acetyl CoA} + 8 \text{ QH}_2 + 8 \text{ NADH} + 8 \text{ H}^+$$

4. In the activation of a fatty acid, one ATP is converted to AMP and PP_i. Pyrophosphate is then hydrolyzed by the action of pyrophosphatase, helping to drive the activation reaction to completion. Therefore, two phosphoanhydride bonds are consumed.

5. The unsaturated intermediate in the β-oxidation pathway has a *trans* double bond between carbons 2 and 3. Natural unsaturated fatty acids usually have *cis* double bonds that are usually not in the same position as the *trans* bonds produced during β-oxidation. Text Figure 16.23 shows the steps necessary for the oxidation of linoleoyl CoA.

6. These ketone bodies are formed when acetyl CoA is present in excess of the amount that can be oxidized by the citric acid cycle. These water-soluble molecules are routed to tissues such as the heart and kidney to supply part of their energy needs. The ketone bodies are more readily transported in the blood than are fatty acids.

7. Bicarbonate adds to acetyl CoA to form malonyl CoA, in which the free carboxylate group was supplied by bicarbonate. When malonyl CoA adds either to an acetyl unit or to the growing fatty acid, the free carboxylate group of malonyl CoA is released as CO_2. Therefore, no label is incorporated into palmitate.

8. Mammals do not have desaturases that act beyond the C-9 position of saturated fatty acids. Thus, the double bonds at C-12 and C-15 cannot be made. Plants do have these desaturases and so can make the C-9, C-12, and C-15 double bonds to produce linolenate.

9. When cholestyramine reaches the intestinal tract, the anionic bile salts are attracted to the positively charged resin and are eliminated from the body. Consequently, LDL is removed from the blood by the liver to release cholesterol, the precursor required for the synthesis of bile salts, to replace those excreted.

10. The ATP yield per carbon atom increases slightly with the length of the carbon chain. Your text indicates that palmitate yields 106 ATP (Section 16.7.D). For stearate, approximately 90 ATP are generated from 9 acetyl CoA, 12 from 8 QH_2, and 20 from 8 NADH, for a total of 122 ATP. Subtracting 2 ATP for activation, the net yield is 120 ATP. The normalized ATP yield for glucose is 32 ATP × 18/6 = 96 ATP. The percentage increase that stearate yields with respect to that of glucose is:

$$(120 - 96)(100)/96 = 25\%.$$

Therefore, the percentage increase is about one percent higher for stearate than it is for palmitate.

11. (a) The citrate transport system operates to export acetyl CoA from mitochondria to the cytosol and to produce NADPH in the cytosol. Both acetyl CoA and NADPH are required for cytosolic fatty acid synthesis.

(b) The system requires energy. Two ATP are required for each acetyl CoA exported to the cytosol.

12. The key regulatory enzyme in fatty acid synthesis is acetyl-CoA carboxylase, which catalyzes the carboxylation of acetyl CoA to form malonyl CoA. The reaction is metabolically irreversible. Citrate activates the enzyme, which is inactivated by phosphorylation. Glucagon stimulates this phosphorylation reaction.

13. Sphingomyelin (see text Figure 16.14) is subject to base-catalyzed hydrolysis and cholesterol is not. Reflux samples of both solutions with aqueous NaOH. Fatty acid salts, which will dissolve in the aqueous phase, will be produced from sphingomyelin. Acidification of the aqueous phase should cause precipitation of the free fatty acids. Cholesterol does not react.

14. Aspirin irreversibly inhibits cyclooxygenase (COX) via acetylation, thus preventing formation of eicosanoids that mediate pain sensitivity and inflammation.

15. Colipase is a small protein that helps bind pancreatic lipase to its lipid substrates in the small intestine. It also activates the lipase by holding the enzyme in a conformation with its active site open.

D. Additional Problems

1. Lack of carbohydrates forces the body to rely on fatty acid oxidation for energy. Supplies of pyruvate and, consequently, oxaloacetate would be inadequate for energy needs due to the lack of carbohydrates. With insufficient oxaloacetate, little acetyl CoA could be consumed in the citric acid cycle. Abnormal quantities of the acetyl CoA would be converted to acetoacetate and β-hydroxybutyrate. Large amounts of these ketone bodies would lead to a drastic lowering of blood pH that could be life-threatening if not checked.

2. Eventually, the supply of CoASH would be depleted by unchecked fatty acid oxidation and energy production would be blocked. Under these conditions, acetoacetate is produced from the accumulating acetyl CoA, a process that releases two CoASH molecules per acetoacetate formed.

3. The LDL-receptor proteins are involved in the removal of LDL from the blood. In individuals with FH, there is essentially no mechanism for removing cholesterol, as LDL, from the blood. Cholesterol derived from LDL also serves to diminish cholesterol synthesis. This control would also be lacking.

4. The protein avidin strongly binds biotin, the prosthetic group of the enzyme acetyl CoA carboxylase. This enzyme catalyzes the carboxylation of acetyl CoA to form malonyl CoA. Since each two-carbon unit added to the growing fatty acid chain is donated by malonyl CoA, prevention of its formation blocks fatty acid synthesis.

5. When fatty acids with odd-number carbon chains are oxidized, the last round of β-oxidation produces propionyl CoA that is converted to succinyl CoA, which can be converted to oxaloacetate. Since oxaloacetate is a substrate for gluconeogenesis, the carbons of the original propionyl CoA molecule are converted to glucose. This is a minor exception to the general rule because fatty acids with an odd-number of carbons are rare in nature.

6. Aspirin is an irreversible inhibitor that acetylates a serine –OH group in the active site of these isozymes. COX-1 regulates secretion of mucin in the stomach that helps protect the gastric wall. It is desirable not to inhibit this isozyme. COX-2 promotes inflammation, pain, and fever. Inhibition of COX-2 would lessen these effects.

Chapter 17

A. True-False

1. <u>False.</u>   Birds eliminate uric acid, aquatic animals eliminate ammonia, and terrestrial invertebrates eliminate urea.

2. <u>False.</u>   Approximately one-half of the 20 amino acids used for protein synthesis in mammals must be obtained in the diet. The number varies among species and with the stage of development.

3. <u>False.</u>   While a portion of the amino acids from dietary proteins is used for cellular protein synthesis, some are deaminated and their carbon skeletons used for other biosynthetic purposes, or to generate energy.

4. <u>True</u>.   L-Glutamate is a substrate that donates an amino group to some α-keto acid acceptor. α-Ketoglutarate is a substrate that serves as an acceptor of an amino group from some amino acid. The transaminases differ in their specificities for particular amino acids and α-keto acids.

5. <u>False</u>.   The liver is the site of urea synthesis.

6. <u>True</u>.   The process of the reduction of one molecule of N_2 to $2 NH_3$ requires a minimum of 16 ATP.

7. <u>True</u>.   As text equation 17.2 shows, one H_2 is produced per each N_2 reduced.

8. <u>True</u>.   This means of carrying NH_4^+ keeps blood levels of this toxic metabolite low.

9. <u>False</u>.   Some amino acids are *both* glucogenic and ketogenic.

10. <u>False</u>.   High levels of arginase are found only in the liver. It catalyzes the hydrolysis of arginine to form ornithine and urea.

11. <u>True</u>.   Bicarbonate is replenished by glutamine catabolism in the kidneys. Each molecule of glutamine furnishes two bicarbonate ions.

12. <u>True</u>.   Nitric oxide is a messenger molecule with several physiological functions. Nitric oxide helps kill bacteria and tumor cells, but is harmful at high concentrations.

13. <u>False</u>.   Some regulatory proteins have very short half-lives because they are specifically targeted for degradation.

14. <u>False</u>.   ATP is required in more than one aspect of the process. (See text Section 17.5.)

B. Short Answers

1. pyridoxal phosphate; to act as a carrier of amino groups

2. lysine; threonine

3. ketogenic; glucogenic

4. phenylketonuria (PKU)

5. atmospheric N_2; soil NO_3^-

6. NH_4^+; aspartate; HCO_3^-

7. carbamoyl phosphate

8. four

9. glucose-alanine cycle

10. the glycine-cleavage system

11. ubiquitin

12. proteasome

13. lysine; methionine; threonine

C. Problems

1. Amino groups are first removed from the amino acids as ammonium ions. The remaining carbon chains of the amino acids are then oxidized for energy by separate pathways.

2. Figure 17.1 shows the reaction, catalyzed by alanine transaminase, in which alanine is converted to pyruvate, which can be converted to oxaloacetate, which can ultimately form glucose by gluconeogenesis. (Note that this reaction is essentially the reverse of that shown in text Section 17.3.C. where the synthesis of alanine is illustrated.)

3.

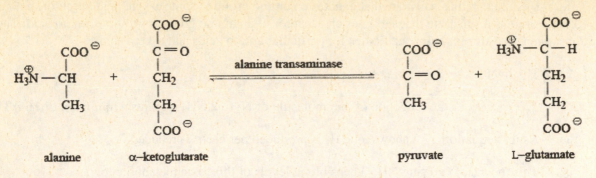

Figure 17.1 Conversion of Alanine to Pyruvate in Liver

4. The positive value of the standard free energy change, +27 kJ/mol, indicates that, under standard conditions (1 M concentrations), the formation of α-ketoglutarate is not spontaneous. In cells, however, concentrations are not 1 M and, further, the NH_4^+ ions produced are constantly being removed and this drives the reaction toward continuous production of α-ketoglutarate.

5. The formation of carbamoyl phosphate requires 2 ATP and releases 2 ADP and P_i as one amino group is brought to the urea cycle. The formation of argininosuccinate from aspartate and citrulline uses one ATP and forms AMP and PP_i. Subsequent hydrolysis of PP_i cleaves a high-energy bond that, together with the previous 3 ATP used, makes a total four high-energy phosphate bonds consumed.

6. The first committed step of the urea cycle, the formation of carbamoyl phosphate, is catalyzed by carbamoyl phosphate synthetase I. N-Acetyl glutamate is an activator of this enzyme. The synthesis of this activator increases as glutamate concentrations increase as a result of transamination reactions involving amino acids being metabolized. The rates of reactions of the urea cycle are controlled by concentrations of their substrates.

7. Text Figure 17.15 illustrates the process by which glycine is produced from serine by the action of serine hydroxymethyltransferase, with pyridoxal phosphate as the prosthetic group. Tetrahydrofolate, derived from folate by reduction, is a coenzyme required for the reaction. Therefore, glycine synthesis is dependent on levels of both serine and folate.

8. Figure 17.2 illustrates the cyclization of glutamate γ-semialdehyde to form a Schiff base that is then reduced to form proline.

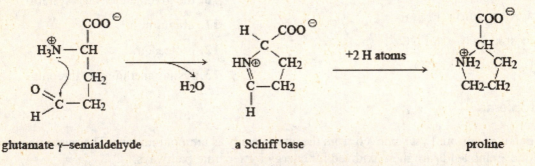

Figure 17.2 Final Steps of Proline Synthesis

9. The seven compounds are oxaloacetate, fumarate, pyruvate, acetyl CoA, α-ketoglutarate, succinyl CoA, and acetoacetyl CoA.

10.

Ketogenic	Glucogenic	Glucogenic-Ketogenic
leucine	alanine	phenylalanine
lysine	cysteine	tyrosine
	glycine	isoleucine
	serine	threonine
	asparagine	tryptophan
	aspartate	
	methionine	
	valine	
	arginine	
	glutamate	
	glutamine	
	histidine	
	proline	

11. Synthesis of tyrosine depends on the presence of the essential amino acid phenylalanine. Cysteine synthesis involves homocysteine, which is derived from the essential amino acid methionine.

12. Caspases are enzymes, activated from their zymogen forms, responsible for hydrolysis of critical cellular compounds that leads to cellular death during the events of apoptosis. (See text Box 17.4.)

13. The sources of the carbon and nitrogen atoms in histidine are glutamate, glutamine, ATP, and phosphoribosylpyrophosphate (PRPP). (See text Figure 17.22.)

14. alanine $+ NAD^+ + H_2O \leftrightarrow$ pyruvate $+ NADH + H^+ + NH_4^+$

(See text section 17.6 and text equation 17.4.)

D. Additional Problems

1. Urease catalyzes the conversion of urea to ammonia and carbon dioxide. The following equation shows that three moles of gas are produced per mole of urea.

$$H_2N-\overset{\overset{\displaystyle O}{\|}}{C}-NH_2 + H_2O \xrightarrow{\text{urease}} 2\,NH_3 + CO_2$$

Mol urea = (2.4 mL gas)(1 mol gas/22.4 × 10^3 mL gas)(1 mol urea/3 mol gas products)
= $3.6 × 10^{-5}$ mol urea.

2. The alanine-glucose cycle operates to remove both pyruvate and NH_4^+ from muscle, in the form of alanine, and to deliver it to the liver, where a reversal of the synthetic reaction releases NH_4^+ and pyruvate. The NH_4^+ is removed by incorporation into urea. Glucose is formed from pyruvate by gluconeogenesis and can be returned to muscle for energy production.

The Cori cycle shuttles lactate, formed by reduction of pyruvate, from muscle to liver, where gluconeogenesis also produces glucose that can be returned to muscle.

The alanine-glucose cycle requires equal amounts of NH_4^+ and pyruvate. If production of pyruvate in muscle exceeds NH_4^+ present, the Cori cycle operates to cycle the excess pyruvate.

Defects in the urea-cycle enzymes can lead to high $[NH_4^+]$.

3. Although the ammonium ion cannot enter the brain, it is in equilibrium with NH_3, which is able to enter the brain. Upon entry, NH_3 becomes protonated to NH_4^+. Parts of the brain utilize glycolysis, the citric acid cycle, and related pathways for energy requirements. NH_4^+ may cause depletion of α-ketoglutarate and glutamate by reacting with them to form glutamate and glutamine, respectively. With a citric acid cycle intermediate in short supply, energy production would be curtailed. Additionally, glutamate is a neurotransmitter and is a precursor for another neurotransmitter, γ-aminobutyrate. Drastic alteration of neurotransmitter levels, as well as reduced levels of energy production, may lead to brain damage.

4. Aspartate is the form in which the second nitrogen atom of urea enters the urea cycle. Aspartate is formed by a transamination reaction in which the amino group from some amino acid is transferred to oxaloacetate.

5. (a) The ^{18}O label is found in AMP in the reaction:

$$Citrulline + Aspartate + ATP \rightarrow AMP + PP_i + argininosuccinate$$

(b) This finding suggests the formation of a citrullyl-AMP intermediate that is subsequently cleaved to yield AMP and argininosuccinate.

Chapter 18

A. True-False

1. <u>False</u>.   The purine fused-ring system contains nine atoms and pyrimidine rings contain six atoms.

2. <u>True</u>.

3. <u>True</u>.   The sources of the various atoms of the purine ring system are illustrated in text Figure 18.2 (formate via 10-formyltetrahydrofolate).

4. <u>False</u>.   The purine ring components are assembled on ribose 5-phosphate. That is, ribose 5-phosphate serves as a foundation upon which the purine ring is built.

5. <u>True</u>.   IMP, which contains the purine base hypoxanthine, is converted to AMP by one branch of the pathway and to GMP by the other branch of the pathway.

6. <u>True</u>.   See text Figure 18.7.

7. <u>True</u>.   See text Figure 18.9.

8. <u>False</u>.   This enzyme is carbamoyl phosphate synthetase II. The urea cycle enzyme is carbamoyl phosphate synthetase I. (See also problem D.1.)

9. <u>False</u>.   Uric acid is a purine.

10. <u>False</u>.   Orotate is an intermediate in the pyrimidine biosynthetic pathway, but uridine 5'-monophosphate (UMP) is the end product of the pathway.

11. <u>False</u>.   The pyrimidine ring system is completed prior to the attachment of ribose 5-phosphate. The attachment occurs in the reaction between orotate and 5-phosphoribosyl 1-pyrophosphate (PRPP). See text Figure 18.10.

12. <u>False</u>.   dTMP is formed by the methylation of dUMP. See text Figure 18.15.

13. <u>False</u>.   Degradation of the purines leads to uric acid, which is the end-product in birds and primates, but there are no distinctive excretory products formed from pyrimidine catabolism.

B. Short Answers

1. purine or pyrimidine base; ribose or deoxyribose; one or more phosphate groups

2. PRPP (5-phosphoribosyl 1-pyrophosphate)

3. the cytosol

4. the smaller (imidazole) ring (containing five atoms)

5. IMP; XMP; GMP; AMP (See text Figure 18.8.)

6. pyrimidines

7. salvage

8. Uric acid

9. glutamine (See text Figure 18.12.)

10. adenine phosphoribosyltransferase; hypoxanthine-guanine phosphoribosyltransferase

11. purine nucleotide cycle (see text Figure 18.22.)

12. acetyl CoA; succinyl CoA (see text Figure 18.23.)

13. allantoin

C. Problems

1. (a) 1 cytosine, 1 ribose

   (b) 1 hypoxanthine, 1 ribose, 1 inorganic phosphate

   (c) 1 guanine, 1 deoxyribose, 3 inorganic phosphates

2. Glutamine furnishes amino groups in steps 1 and 4 of the biosynthesis of IMP from PRPP, and also in the formation of GMP from XMP. The nitrogens of the donated groups become nitrogen atoms 9 and 3, respectively, of the purine ring, and the 2-amino group of GMP. (See text Figures 18.5 and 18.7.)

3. PRPP synthetase has been shown to be inhibited by several purine ribonucleotides, but not at concentrations usually present in cells. The major point of control of purine biosynthesis is discussed in a question to follow.

4. Text Figure 18.3 shows that one ATP is used to form PRPP, but two high-energy bonds are consumed since the pyrophosphate group of PRPP is removed in step 1 of the *de novo* synthesis and is hydrolyzed to $2 P_i$. Text Figure 18.5 shows that four other ATP molecules are used (steps 2, 4, 5, 7) and one high-energy bond of each is consumed. Text Figure 18.7 shows that one GTP is required and one of its high-energy bonds is consumed. This gives a total of seven ATP equivalents (high-energy bonds).

5. (a) In such cases, it is likely that the allosteric enzyme does have more than one binding site for the inhibitors. This enzyme has separate sites for the binding of guanine and adenine nucleotides.

   (b) It is also likely that complete inhibition is not achieved by the binding of any one single inhibitor. It is more likely that the greatest inhibition is caused by the binding of all of the inhibitors.

6. $2 ATP + glutamine + 2 H_2O + Q + PRPP \rightarrow 2 ADP + 4 P_i + glutamate + QH_2 + UMP$

7. Aspartate condenses with carbamoyl phosphate to form carbamoyl aspartate in step 2 of the pathway illustrated in text Figure 18.10. The –COOH group is retained until step 6 in which it is released as HCO_3^- as OMP is converted to UMP.

8. Unlike most enzymes that catalyze carboxylation reactions, which require ATP and the prosthetic group biotin, AIR carboxylase requires neither biotin nor ATP. The mechanism for this carboxylation is shown in text Figure 18.6.

9. Figure 18.1 illustrates the formation of CTP from UTP.

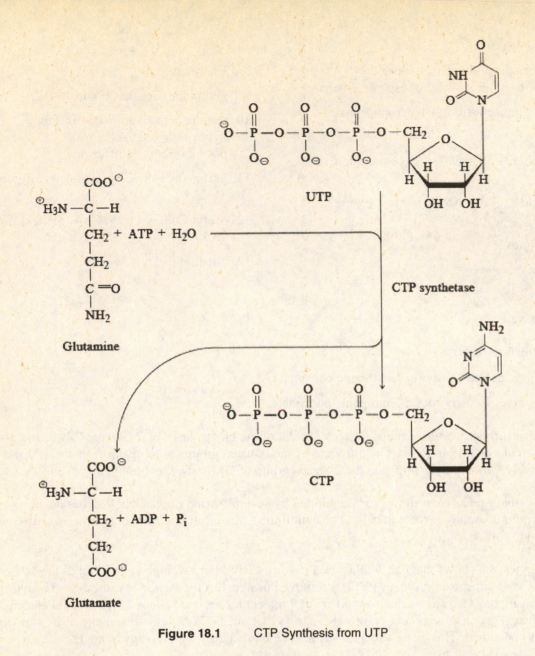

Figure 18.1   CTP Synthesis from UTP

10. The label will appear in the 5-methyl group of dTMP. (See text Figure 18.15.)

11. Deoxyribonucleotides are synthesized by the reduction of ribonucleotides. An enzyme, ribonucleoside diphosphate reductase, catalyzes the reduction of all four ribonucleoside diphosphates (ADP, GDP, CDP, UDP). Reducing power is supplied by NADPH and is transferred to the ribose moiety by the involvement of FAD, thioredoxin, and sulfhydryl groups of the enzyme.

12. (a) Lesch-Nyhan syndrome results from a hereditary deficiency of hypoxanthine-guanine phosphoribosyltransferase, a salvage enzyme that converts hypoxanthine and guanine to IMP and GMP, respectively. If not so salvaged, free hypoxanthine and guanine are degraded to uric acid.

(b) The disease is generally restricted to males because the gene that codes for the deficient enzyme is located on the X chromosome.

(c) Symptoms include mental retardation, palsy, and self-mutilation.

(d) Since much of the hypoxanthine and guanine normally salvaged is degraded to uric acid, increased *de novo* synthesis is likely to result from decreased inhibition of glutamine-PRPP amidotransferase by IMP and GMP.

D. Additional Problems

1. Carbamoyl phosphate synthetase I is found in liver mitochondria, but carbamoyl phosphate synthetase II is a cytosolic enzyme. Compartmentalization allows the latter enzyme to be allosterically controlled without an effect on the mitochondrial enzyme involved in urea synthesis.

2. Glutamine furnishes N-3 and HCO_3^- furnishes C-2 of the pyrimidine ring. (See text Figure 18.9.) The CO_2 eliminated when OMP is converted to UMP was supplied by aspartate. (See text Figure 18.10.) Therefore, both OMP and UMP should contain the ^{14}C-label, but CO_2 should not.

3. In the previous chapter of this study guide, problems 17 II.B.8. and 17 II.C.4. showed that four ATP are required for the synthesis of one molecule of urea. Incorporation of one NH_4^+ into urea then requires two ATP. One GTP is required in the reaction of IMP and aspartate to produce adenylosuccinate and GDP and P_i. Therefore, the energy requirement to produce one fumarate by the purine nucleotide cycle, and the elimination of NH_4^+, is the equivalent of three ATP molecules.

4. dUTPase catalyses the hydrolysis of dUTP to dUMP and PP_i. This action prevents dUTP from being incorporated into DNA in place of dTTP. Incorporation of dUTP in any quantity into DNA would interfere with replication and transcription. Cellular death would be the ultimate effect.

5. Increased availability of glucose 6-phosphate, due to the deficiency in glucose 6-phosphatase, stimulates the pentose phosphate pathway in the liver. Increased production of ribose 5-phosphate leads to increased amounts of PRPP, which stimulate purine biosynthesis. The latter leads to higher levels of uric acid and, hence, to gout.

6. (a) In the biosynthetic pathway for IMP, the enzymes of steps 1 and 4, glutamine-PRPP amidotransferase and FGAM synthetase would be inhibited since they both utilize glutamine as a substrate. GMP synthetase, which catalyzes the glutamine-dependent conversion of XMP to GMP, would be inhibited by azaserine, as would carbamoyl phosphate synthetase in the UMP biosynthetic pathway. Other enzymes including CTP synthetase would also be inhibited. (See text Figures 18.5, 18.7, 18.10, and 18.12.)

(b) Affinity labels bind covalently to active site groups, causing irreversible inhibition. Azaserine is known to covalently bind to the sulfhydryl group of a cysteinyl residue in the active sites of several of these glutamine-dependent enzymes. Figure 18.2 illustrates this kind of inhibition.

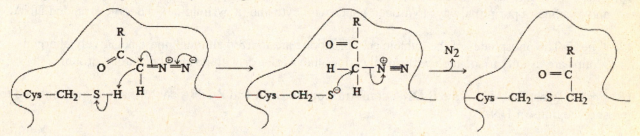

Figure 18.2 Alkylation of an Active-Site Cysteine by a Diazo-antibiotic such as Azaserine

7. Cancer tissues rapidly divide and require DNA for this cell division. 5-Fluorouracil is converted in cells to 5-fluorodeoxyuridylate, which tightly binds thymidylate synthase, thus preventing formation of TMP needed (as TTP) for DNA formation. Without adequate DNA, the cancer tissue cannot survive.

Chapter 19

A. True-False

1. <u>False</u>.   Nucleotides contain three different kinds of components: a sugar (ribose or deoxyribose), a nitrogenous base (a purine or pyrimidine), and one or more phosphate groups.

2. <u>True</u>.   The pentose D-ribose is derived from RNA and 2-deoxy-D-ribose is derived from DNA.

3. <u>False</u>.   Uracil and thymine are pyrimidines, but uracil is a component of RNA, not DNA.

4. <u>True</u>.

5. <u>True</u>.   Miescher isolated the material in 1869.

6. <u>False</u>.   %A = %T and %G = %C, but %(A + T) can be any value (technically) between 0 and 100%. In naturally occurring DNA samples, the %(A + T) ranges from about 30% to about 70%. (See Figure 19.1.)

7. <u>False</u>.   The Watson-Crick model of DNA is an *antiparallel*, double-stranded, right-handed helix.

8. <u>False</u>.   The separated DNA strands absorb *more* UV radiation because the bases are deshielded.

9. <u>True</u>.   These enzymes are not restricted to cleaving terminal phosphodiester bonds as are exonucleases.

10. <u>True</u>.   While the majority of DNA in eukaryotic cells (that in the nucleus) is not circular, DNA in mitochondria and chloroplasts is circular. In general, bacterial DNA is also circular.

11. <u>False</u>.   DNA is not subject to base-catalyzed hydrolysis as RNA is.

12. <u>True</u>.   See text Section 19.6.A and B.

13. <u>True</u>.   The restriction endonucleases, however, will not catalyze hydrolysis of the DNA strand if the recognition sequence is methylated.

14. <u>False</u>.   Sticky ends are formed when restriction endonucleases catalyze cleavage of the two strands of DNA at different places. This leaves duplex DNA with one strand at each cleavage site longer than the other. The DNA is not denatured.

15. <u>False</u>.   By convention, the genome of a species does not include the mitochondrial or chloroplast DNA.

16. <u>True</u>.   The ratio is 1:1 because adenine (a purine) always hydrogen bonds with thymine (a pyrimidine) and guanine (a purine) always hydrogen bonds with cytosine (a pyrimidine) in double-stranded DNA.

17. <u>True</u>.   The phosphates of the nucleic acid backbone are ionized although these negatively charged groups are usually associated with small cations like Mg^{2+} or with positively charged proteins.

18. <u>True</u>.   Most DNA is of the B-DNA conformation. A-DNA and Z-DNA conformations are confined to short regions of DNA.

B. Short Answers

1. uracil

2. cytosine

3. dTTP

4. deoxynucleoside triphosphates (dTTP, dATP, dGTP, dCTP). The corresponding monophosphate residues are actually incorporated in the DNA chain.

5. B-DNA

6. Z-DNA

7. adenine; guanine

8. melting (thermal denaturation); T_m, the melting temperature

9. restriction endonucleases

10. restriction sites (Many restriction sites are palindromic.)

11. histones; nucleosome

12. hydrogen bonds; charge-charge interactions; hydrophobic effects; stacking interactions (London dispersion forces)

13. topoisomerases

14. messenger RNA (mRNA)

15. S-adenosylmethionine

16. eukaryotic

C. Problems

1. ATP and other related nucleotide triphosphates are ionized at cellular pH and apparently form complexes with Mg^{2+} ions in cells. Specifically a Mg^{2+} ion complexes with the second and third phosphate groups of ATP, forming a β,γ complex.

2. The cytosine content is the same (17%) as the guanine content since they form complementary base pairs. Adenine forms complementary base pairs with thymine and the total of adenine + thymine = 100 - 2(17%) = 66%. Therefore, the %adenine = %thymine = 33%.

3. The average rise in a DNA chain (distance between bases) is 0.33 nm. The number of bases will be:

    bases = 1 base pair/0.33 nm × 51 nm × 2 bases/pair
    = 315 or 320 bases (to two significant figures).

4. The base sequence C-G-C-G-C-G is complementary to itself in the opposite direction.

    5'–C–G–C–G–C–G–3'
    3'–G–C–G–C–G–C–5'

5. Figure 19.1 shows that a T_m of 90 °C corresponds to a G + C content of about 51 mol%, which means that the A + T content is 49 mol%, and the adenine is 24.5 mol%.

6. The T_m would be higher than that of the DNA in the previous problem. With a lower percentage of A-T pairs, there would be more G-C pairs, thus the stacking interactions would be stronger and there would be more hydrogen bonds and a higher T_m. From Figure 19.1, the T_m is about 99°C.

7. Some bacteria have both restriction methylases and restriction endonucleases that recognize the same sequences in DNA. The methylases catalyze the addition of methyl groups to these sequences of the cell's own DNA. This serves as a protection and prevents the cell's restriction endonucleases from cleaving its own DNA. Unmethylated foreign DNA is unprotected and is cleaved by the bacterial cell's restriction endonucleases. (Also note that hemimethylated sites of newly replicated DNA are high-affinity substrates for the methylase, but are not recognized by the restriction endonuclease. This assures continued protection of the cell's own DNA.)

8. As complementary bases form pairs, the pairs are between the two chains. The base pairs are stacked above and below each other between the two chains. Cooperative interactions between the base pairs

bring the pairs closer together and cause the sugar-phosphate backbone to twist into the helix shape. These stacking interactions are responsible, in part, for the helix formation.

9. Ribosomal RNA, rRNA, makes up about 80% of all cellular RNA. It is a component of the cytoplasmic particles called ribosomes, which are involved in protein synthesis.

10. Nucleosomes are composed of histones and DNA. A histone octamer, called a core particle, forms from two each of histones H2A, H2B, H3, and H4. About 146 base pairs of DNA wrap around the core particle, with a total of 200 base pairs being involved in the formation of a nucleosome. An H1 histone can also be associated with the linker DNA between nucleosomes. Coiling of the beads-on-a-string assembly of nucleosomes gives rise to a more condensed structure called the 30-nm fiber.

11. *Hae*III and *Sma*I catalyze nonstaggered cuts, producing blunt ends. These enzymes catalyze the cleavage of both strands of duplex DNA at the same location so that no single-stranded segments, or sticky ends, are formed.

D. Additional Problems

1. (a) Since there are 200 base pairs per nucleosome, the total number of nucleosomes in the chromosome would be:

$$(2.4 \times 10^8 \text{ bp})(1 \text{ nucleosome}/200 \text{ bp}) = 1.2 \times 10^6 \text{ nucleosomes.}$$

(b) The average number of nucleosomes per loop of this condensed DNA would be:

$$(1.2 \times 10^6 \text{ nucleosomes})/(2000 \text{ loops}) = 6.0 \times 10^2 \text{ or } 600 \text{ nucleosomes/loop.}$$

2. Since the exonucleases do not act on this DNA, it may be circular. In double-stranded DNA, the thymine/adenine ratio is close to 1.00. The ratio 1.36 suggests that normal A-T pairing is not present. One explanation is that the DNA is not only circular, but that it is also single-stranded.

3. Circular DNA can exist in relaxed circles or in supercoiled forms. (See text Figure 19.19.) Supercoiled circular DNA is more compact than the same DNA in relaxed circles. Vinograd's DNA contained some of both relaxed and supercoiled forms. The more compact, supercoiled DNA sedimented more quickly and the relaxed form of the circular DNA sedimented less quickly, giving rise to the two DNA bands observed.

4. The somewhat random action of DNase I suggests that no specific base sequence in the substrate DNA is required by the enzyme. Lysine and arginine residues of the enzyme are positively charged and participate in charge-charge interactions with negatively charged phosphate groups of the sugar-phosphate backbone of B-DNA. This is thought to occur in the minor groove of B-DNA, but cannot occur in Z-DNA because its minor groove is too narrow for enzyme access. Also, the sugar-phosphate backbone of Z-DNA is oriented more toward the interior of the molecule than is the case in B-DNA.

5. Cytosine is converted to uracil by nitrous acid and adenine is converted to hypoxanthine, as shown in Figure 19.2. Uracil does not hydrogen bond properly with guanine, the normal base pair partner of cytosine. Likewise, hypoxanthine does not hydrogen bond properly with thymine, the normal partner of adenine. These changes would cause errors in replication and transcription.

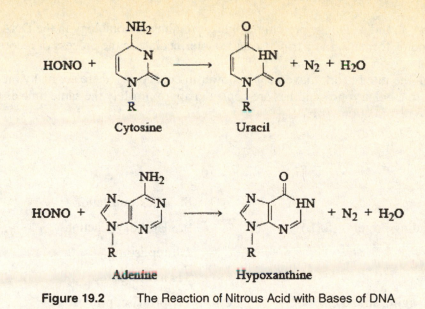

Figure 19.2 The Reaction of Nitrous Acid with Bases of DNA

Chapter 20

A. True-False

1. <u>False</u>. The *semiconservative* model of DNA replication is supported by the experiments of Meselson and Stahl and others.

2. <u>False</u>. Replication of the circular, duplex DNA of *E. coli* occurs in a bidirectional manner from a single initiation site.

3. <u>False</u>. Not the deoxyribonucleoside monophosphates, but the triphosphates (dATP, dGTP, dCTP, dTTP) are required. Pyrophosphate, PP_i, is released from each deoxyribonucleoside triphosphate during incorporation of the nucleotide monophosphate moiety into the DNA strand.

4. <u>True</u>. *E. coli* cells have three different DNA polymerases and eukaryotes have at least five different DNA polymerases.

5. <u>False</u>. DNA polymerase I also requires pre-formed DNA to serve both as a primer strand and as a template strand.

6. <u>True</u>. DNA polymerase III catalyzes chain elongation during DNA replication. It is a key component of the replisome.

7. <u>False</u>. DNA polymerase III requires both a template and a primer (a 3' end of a pre-existing nucleotide chain).

8. <u>False</u>. It is essentially an irreversible process due to the hydrolysis of released PP_i, catalyzed by pyrophosphatase.

9. <u>True</u>. Eukaryotic Okazaki fragments contain 100–200 nucleotide residues, but prokaryotic Okazaki fragments contain about 1000 nucleotide residues.

10. <u>True</u>. DNA ligase catalyzes the joining of one strand segment to another (seals nicks) by catalyzing the formation of phosphodiester bonds.

11. <u>False</u>. Thymine dimers are corrected by the action of DNA photolyase. The process is called photoreactivation and is a type of direct repair, not excision repair.

12. <u>True</u>.   The RecA strand exchange protein recognizes regions of homology in the DNA single and duplex strands and promotes formation of a triple-stranded intermediate in the process of recombination.

13. <u>False</u>.   Although the rate of fork movement is slower in eukaryotes, there are many independent origins of replication so that eukaryotic genomes are copied in approximately the same time as those of prokaryotes. (See text Section 20.1.)

B. Short Answers

1. replication fork
2. single-strand binding proteins (SSB)
3. template
4. primers
5. leading strand; lagging strand
6. processive
7. Okazaki fragments
8. Nick translation
9. Holliday junction
10. replisome
11. helicases
12. primase; primosome
13. NAD^+
14. terminator utilization substance (Tus)

C. Problems

1. Circular duplex DNA in the bacterial cell forms large loops and is highly supercoiled. It is, therefore, very compact.

2. The phosphate groups in DNA are ionized under cellular conditions, carry negative charges, and make the exterior of the helical molecule highly polar. Acid furnishes H^+ ions that protonate the phosphate anions, rendering them nonionic and making the exterior much less polar and, therefore, less soluble in water.

3. Each new strand of DNA is synthesized in a 5'→3' direction.

4. The proofreading function of DNA polymerase III involves the 3'→5' exonuclease activity of the enzyme, which catalyzes the removal of a mismatched deoxyribonucleotide that has just been incorporated into a new DNA strand during replication.

5. As depicted in text Figure 20.15, there are two polymerase core complexes, one for each of the DNA template strands. The lagging strand template loops back through the replisome so that synthesis of both DNA strands occurs in the direction of movement of the replication fork.

6. In the final stages of lagging-strand synthesis, RNA primers are removed by the 5'→3' exonuclease activity of DNA polymerase I, which then performs nick translation to fill in the gaps left by the RNA primers. Once these new DNA segments are in place, the sugar-phosphate backbone is sealed by the action of DNA ligase. Therefore, DNA samples isolated after replication contain no RNA.

7. 2',3'-Dideoxyribonucleoside triphosphates are nucleoside triphosphates that lack oxygen at both the 2' and 3' positions on the ribose moiety. During DNA synthesis, they are unable to form a phosphodiester link at the 3' position as are normal deoxyribonucleoside triphosphates. Incorporation of a dideoxy form terminates chain synthesis. ddNTPs are used in the Sanger method of DNA sequencing, as discussed in text Box 20.1.

8. Because DNA replication is processive, fewer enzyme molecules are needed and a more rapid rate is realized for the overall process.

9. The damaged region is detected and an endonuclease cleaves the sugar-phosphate backbone on both sides of the damaged area, so that the oligonucleotide is then removed. The resulting gap is filled in by the

action of DNA polymerase I or a similar enzyme and the nick in the sugar phosphate backbone is sealed by the action of DNA ligase. (See text Section 20.7.B and text Figure 20.21.)

10. DNA is replicated in the nucleus of eukaryotic cells and proteins are synthesized in the cytosol. As indicated, the number of histones doubles with each round of replication. The newly formed histones are used in the formation of nucleosome and chromatin structures of daughter DNA molecules.

11. The *E. coli* chromosome contains 4.6×10^6 nucleotide pairs or 9.2×10^6 total bases, all of which must be replicated. One error occurs per 10^7 replication steps. The average number of incorrect bases is:

9.2×10^6 polymerization steps $\times$ 1 error/10^7 polymerization steps = 0.92 incorrect bases.

About one incorrect base is incorporated per replication of the chromosome.

12. (a) Both DNA ligases form an AMP-DNA-ligase intermediate. The *E. coli* DNA ligase uses NAD^+ as cosubstrate and supplier of the AMP moiety, while eukaryotic DNA ligase uses ATP for this function.

(b) The kinetic mechanism is ping-pong. The cosubstrate (NAD^+ or ATP) binds first to the enzyme and the first product (NMN^+ or PP_i) is then released before the next substrate (the nicked DNA) binds and reacts.

(c) The DNA ligases do not have a specificity for any particular base sequence in DNA. They recognize the gap between the 3'-OH group of one ribose and the adjacent 5'-phosphoryl group. (Note, in the caption of text Figure 20.14, "B" can be any of the bases found in DNA.)

D. Additional Problems

1. Gel electrophoresis of two such samples could help to identify the prokaryotic Okazaki fragments as being the longer, higher molecular weight fragments. The eukaryotic fragments would travel farther down the gel since they are typically only one- to two-tenths the size of prokaryotic Okazaki fragments.

2. (a) There are 4.6×10^6 nucleotide pairs in the *E. coli* chromosome, but only one DNA strand (4.6×10^6 nucleotides) is replicated discontinuously to produce Okazaki fragments. Each Okazaki fragment (OF) is 1000 nucleotides in length. The number of fragments in one-third of the chromosome is:

$$4.6 \times 10^6 \text{ nucleotides} \times 1 \text{ OF}/10^3 \text{ nucleotides} \times 1/3 = 1530 \text{ or about } 1500 \text{ OF.}$$

(b) The rate of new strand extension is 1000 nucleotides per second (text Section 20.1.). Nucleotides are incorporated simultaneously along the leading strand and the lagging strand at this rate. Replication is bidirectional, therefore, there are two replication forks moving in opposite directions on the circular DNA molecule making the overall rate 2×10^3 nucleotides per second. The time is:

$(4.6 \times 10^6$ nucleotides/chromosome)(1 s/2×10^3 nucleotides)(1/3 chromosome)(1 min/60 s) = 12.7 minutes.

3. (a) The ^{3}H-labeled DNA is newly synthesized DNA that was formed by incorporation of the ^{3}H-deoxythymidine during replication. (The ^{3}H-deoxythymidine was first converted to ^{3}H-TTP by the cell.)

(b) Both short and long fragments of newly synthesized DNA occurred because of the different modes of synthesis on the two existing DNA strands that serve as templates. Discontinuous synthesis on the lagging strand produced the short DNA fragments. (See text Figure 20.10.)

4. The direction of travel of the fragments in the gel shown in Figure 20.1 is from top to bottom. The smallest fragments travel the farthest down the gel. The smallest fragment will be that formed by the attachment of one base (the ddNTP) to the primer. The third lane contains the smallest fragment, which must be 5'-TAC-3', since ddCTP was used in that trial. Using this reasoning, the sequence of bases that attach to the primer can be read from bottom to top of the gel. That sequence, including the primer, is 5'-*TACCAATGGC*-3'. The complementary DNA strand, which served as a template for synthesis of the

strand attached to the primer, must be 5'-GCCATTGGTA-3'. The duplex DNA segment is shown here with the primer in italics.

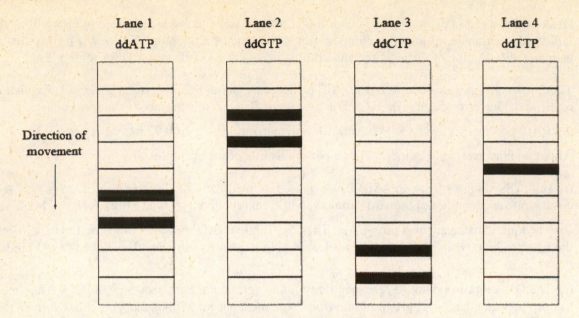

Figure 20.1 Sanger's DNA Sequencing Method

5'-*TA*CCAATGGC-3' (synthesized strand)
3'-ATGGTTACCG-5' (existing strand)

5. The fragments are listed from smallest to the largest in each lane.

> *Lane 1*: 5'-TACCA-3'; 5'-TACCAA-3'
>
> *Lane 2*: 5'-TACCAATG-3'; 5'-TACCAATGG-3'
>
> *Lane 3*: 5'-TAC-3'; 5'-TACC-3'; 5'-TACCAATGGC-3'
>
> *Lane 4*: 5'-TACCAAT-3'

Chapter 21

A. True-False

1. <u>True</u>.   Synthesis of the RNA polymer is in a 5'→3' direction.

2. <u>False</u>.   Modification of the four common bases occurs after polymerization (i.e., transcription) to produce these unusual bases.

3. <u>False</u>.   Promoters are double-stranded DNA sequences. RNA polymerase binds to both strands during transcription initiation.

4. <u>True</u>.   The DNA template strand is complementary to the DNA coding strand. The RNA nucleotides incorporated alongside the DNA template strand are complementary to those of the DNA template strand, which makes the RNA base sequence identical to that of the DNA coding strand, except that U replaces T.

5. <u>True</u>.   The housekeeping genes are said to be constitutively expressed. They are not inducible.

6. <u>True</u>.   Translation is coupled to transcription in prokaryotes.

7. <u>True</u>.　The poly A tails on the 3' end help prevent exonuclease cleavage of the coding region of the mature mRNA.

8. <u>False</u>.　This cap is added to the 5' end of the mRNA shortly after it is synthesized. The poly A tail is added to the 3' end of the mRNA.

9. <u>True</u>.　During processing, spacer sequences are removed from the large precursor rRNA molecule to yield the individual rRNA molecules that are components of eukaryotic ribosomes.

10. <u>True</u>.　The σ^{70} subunit is mainly responsible for formation of an initiation complex at the site of σ^{70}-specific promoters.

11. <u>False</u>.　Three RNA polymerase complexes catalyze transcription in the cell nucleus, but different RNA polymerase complexes are also found in mitochondria and in chloroplasts of those eukaryotes that contain mitochondria and chloroplasts. (See text Table 21.4 for more details.)

12. <u>True</u>.　The process of splicing is catalyzed by a large RNA-protein complex called the spliceosome.

13. <u>False</u>.　There is rapid turnover of mRNA. An *E. coli* cell devotes about one-third of its transcription activity to mRNA synthesis.

14. <u>False</u>.　Small RNA molecules of the fourth class of RNA do have catalytic activity.

15. <u>False</u>.　CRP-cAMP interacts with DNA sequences near promoters of more than 30 genes, of which the lac operon is one.

B. Short Answers

1. transfer RNA; ribosomal RNA; messenger RNA; small RNA

2. ribosomal RNA

3. promoter

4. RNA polymerase holoenzyme (Note that the holoenzyme includes the sigma factor which dissociates from the holoenzyme after transcription is initiated. The core polymerase, consisting of the $\alpha_2\beta\beta'$ subunits, catalyzes elongation of the RNA molecule.)

5. transcription bubble

6. rho Factor

7. RNA polymerase II

8. introns; exons (These terms also refer to the regions of the gene (DNA) than encode corresponding RNA introns and exons.)

9. small nuclear RNA's (snRNA's)

10. repressor

11. activator

12. inducers

13. housekeeping genes

C. Problems

1. In *E. coli*, the rate of transcription is from 30 to 85 nucleotides per second. DNA synthesis occurs about 10 times faster.

2. The TATA box is a region of consensus sequence, TATAAT, that is part of some promoters. It is part of the DNA sequence to which RNA polymerase binds during the initiation of transcription.

3. The protein rho binds to single-stranded RNA behind the paused transcription complex, and promotes disruption of the transcription complex and its dissociation from the DNA template strand.

4. The *lacI* gene is located upstream from the *lac* operon. It encodes a regulatory protein called *lac* repressor that binds to DNA near the promoter of the *lac* operon and prevents transcription.

5. In catabolite repression, one catabolite represses the expression of enzymes required for the catabolism of other catabolites. For example, in a growth medium that contains glucose along with other catabolites (for example, lactose), *E. coli* preferentially uses glucose. In this case, glucose causes catabolite repression. In the absence of glucose, levels of cAMP rise resulting in the activation of genes that are normally not transcribed at high rates in the presence of glucose.

6. CRP (cAMP regulatory protein) is a prokaryotic transcriptional regulatory protein. It binds cAMP to form a complex that can act as an activator of genes subject to catabolite repression.

7. *RNA polymerase I* catalyzes the synthesis of large precursor rRNA molecules.

RNA polymerase II catalyzes the synthesis of precursor mRNA molecules and transcribes some genes that encode small RNA molecules.

RNA polymerase III catalyzes the synthesis of tRNA, 5S rRNA, and some other small RNA molecules.

8. Capping of eukaryotic mRNA molecules at their 5' ends with 7-methylguanylate groups protects the mRNA molecules from degradation by exonucleases. It also promotes proper binding of mRNA molecules to ribosomes in preparation for protein synthesis.

9. Text Section 21.2.B indicates that *E. coli* transcription rates range from 30 to 85 nucleotides per second. Selecting an average of 58 nucleotides per second, the time required would be:

$(4.6 \times 10^6 \text{ bp})(0.978) \times 1 \text{ s}/58 \text{ bp} \times 1 \text{ hr}/3600 \text{ s} = 21.5 \text{ hr}$, or about one day for a single RNA polymerase molecule to transcribe 97.8% of the *E. coli* chromosome.

10. (a) The error rate in RNA synthesis is higher than that of DNA replication because RNA polymerase does not have the exonuclease proofreading activity possessed by DNA polymerase III.

(b) In some cases (sickle cell anemia), a single amino acid substitution in a protein can be disadvantageous. If the error is in the DNA, it affects all molecules of the encoded protein. If an error occurs in synthesizing an mRNA molecule, the problem resides in only that one molecule and not in other copies of the mRNA, so only a small fraction of protein molecules are defective.

11. In regions of dyad symmetry, a segment of one DNA strand contains the same sequence of bases as a nearby segment in the other DNA strand. These segments are not hydrogen bonded to each other, but are separated in the duplex DNA by a short segment (e.g., four base pairs, as shown in text Figure 21.8). Such a region aids or exaggerates a pause in transcription because the transcript RNA molecule forms a hairpin that may destabilize the DNA-RNA transcription hybrid.

12. The consequence is that expressed genes must be continuously transcribed and that most mRNA molecules are translated only a few times.

D. Additional Problems

1. Figure 21.1 illustrates the induction of the *lac* operon by IPTG since *lac* mRNA begins to appear after IPTG is added to the medium. The operon is transcribed at a high rate in the absence of glucose because CRP-cAMP is present. The addition of glucose lowers cAMP levels, thus abolishing activation by CRP-cAMP. This illustrates catabolite repression.

2. The original concentration of R is

$(10 \text{ molecules/cell}) \times (1 \text{ mol}/6.02 \times 10^{23} \text{ molecules}) \times (1 \text{ cell}/0.30 \times 10^{-15} \text{ L}) = 5.5 \times 10^{-8} \text{ M}.$

The original concentration of O is

$$(2 \text{ molecules/cell}) \times (1 \text{ mol}/6.02 \times 10^{23} \text{ molecules}) \times 1 \text{ cell}/0.30 \times 10^{-15} \text{ L}) = 1.1 \times 10^{-8} \text{ M}.$$

At equilibrium, $K_D = [R][O]/[RO] = 10^{-13}$ M. Practically all O is complexed due to the high association constant ($K_{Assoc} = 10^{13}$ M^{-1}). Thus, we can estimate the concentration of RO as 1.1×10^{-8} M. The equilibrium concentration of R will be 5.5×10^{-8} M $- 1.1 \times 10^{-8}$ M $= 4.4 \times 10^{-8}$ M. Then,

$$[O] = K_D[RO]/[R] = (10^{-13})(1.1 \times 10^{-8})/(4.4 \times 10^{-8}) = 2.5 \times 10^{-14} \text{ M}.$$

Compared with the original concentration of O, 99.9998% of O is bound by R at equilibrium.

3. The data in Figure 21.2 show that three components of the mixture were detected by their radioactivity. Most of the RNA was found in the first fractions, which were more dense and sedimented toward the bottom of the CsCl gradient. The second RNA fraction occurs at about the same location as does the majority of the less dense, denatured DNA. This was taken as evidence of a DNA-RNA hybrid that should form if their sequences were complementary, as they would be if one DNA strand had served as a template for synthesis of the RNA. This fraction containing the hybrid does not occur between that of the nonhybridized RNA and DNA because the RNA segment in the hybrid is short compared to the length of the DNA strand with which it hydrogen bonds. Although not shown in Figure 21.2, no DNA-RNA hybrid was formed unless the DNA was first denatured.

4. After about 10 nucleotides are incorporated into the RNA chain, the holoenzyme undergoes a conformational change that causes release of the σ^{70} subunit. The core polymerase no longer binds the promoter sequence any more strongly than other DNA sequences and is able to move away from the promoter (promoter clearance). Accessory proteins, such as Nus A, may also be involved in this process.

5. Not removing an intron region from an mRNA precursor would produce an aberrant mRNA with a disrupted coding frame. The resulting protein might contain additional amino acids or translation might terminate prematurely due to the presence of in-frame stop codons in the intron.

Chapter 22

A. True-False

1. <u>True</u>. There must be at least one specific tRNA for each of the 20 amino acids of which proteins are made.

2. <u>True</u>. Typically, tRNA molecules range from 73 to 95 nucleotides in length.

3. <u>False</u>. The aminoacyl-tRNA synthetases bind specific tRNA's as well.

4. <u>False</u>. Text Figure 22.9 shows the formation of an aminoacyl adenylate as PP$_i$ is released from ATP. While subsequent hydrolysis of the PPi product helps drive the reaction, the AMP moiety of the ATP serves as a temporary carrier of the amino acid.

5. <u>True</u>. This is also true in eukaryotic systems.

6. <u>False</u>. The amino acids are covalently attached, ultimately, to the 3' end of the tRNA. Initially, they may be attached to the 2'- or 3'-hydroxyl group of the adenylate residue of the terminal CCA sequence depending on whether a Class I or Class II aminoacyl-tRNA synthetase was involved.

7. <u>False</u>. It does so in many prokaryotes, but in eukaryotes, methionine is not formylated.

8. <u>True</u>. See text Figure 22.8.

9. <u>True</u>. The Shine-Delgarno sequence, located upstream of the initiator codon, is the ribosomal binding site.

10. <u>True</u>.   See text Section 22.1.

11. <u>False</u>.   The peptidyl transferase activity is located in the 50S ribosomal subunit and appears to be localized in the 23S rRNA molecule of the ribosomal subunit. (RNA molecules that catalyze reactions are called ribozymes by some and RNA enzymes by others.)

12. <u>True</u>.   Since the peptide chain grows from the amino toward the carboxy terminus, the amino group of the new amino acid being incorporated must be free to react.

13. <u>False</u>.   Prokaryotic mRNA molecules may be polycistronic, but those of eukaryotes generally are not.

14. <u>True</u>.   Eukaryotic mRNA molecules do not have the Shine-Delgarno sequences, but they do have the 5'-cap that helps in the binding of the small ribosomal unit.

15. <u>False</u>.   Such proteins are transported into the lumen of the endoplasmic reticulum, then to the Golgi apparatus, and then into vesicles for transport to the plasma membrane for secretion. Posttranslational modification of these proteins can occur in each of these three locations.

B. Short Answers

1. methionine; tryptophan

2. Isoacceptor tRNA molecules

3. phenylalanine; UUU

4. proofreading

5. acceptor stem

6. synonymous codons

7. translocation

8. polysome (or polyribosome)

9. four

10. signal peptide

11. release factors

12. cotranslational; posttranslational

C. Problems

1. An amino acid is made ready for incorporation into a protein by its covalent attachment to a tRNA molecule specific for that amino acid. An aminoacyl adenylate intermediate reacts with the 3'-OH group of the terminal adenylate group of the tRNA to form an ester linkage. The reaction is catalyzed by an aminoacyl-tRNA synthetase specific for both the amino acid and the tRNA molecule. The reactions are outlined in text Figure 22.9.

2. The Class I synthetases aminoacylate tRNA at the 2'-OH group, while the Class II synthetases aminoacylate tRNA at the 3'-OH group.

3. No, the error is only 1 in 10,000 due to the proofreading activity of the isoleucyl-tRNA synthetase. (See text Section 22.3.C.)

4. No bacterial mRNA was being produced in this system, since there were no intact cells. Since mRNA is short-lived, there probably was very little in the supernatant fraction. This experiment showed that mRNA provided the templates necessary for protein synthesis.

5. Assuming a triplet code, one could conclude:
    (a) A code for lysine is AAA.

    (b) A code for asparagine is AAC.

    (c) mRNA templates are translated in a 5'→3' direction.

    (d) Proteins are synthesized from their amino terminus toward their carboxy terminus.

6. Yes, this allows the bacterial cell to start synthesis of proteins encoded by a particular mRNA molecule even before synthesis of the mRNA molecule is complete.

7. The term degenerate refers to the fact that there is more than one codon for most amino acids. Synonymous codons are those that encode the same amino acid.

8. (a) If the amino acid was recognized by the codon, the radioactive alanine would have been incorporated into some of the hemoglobin sites normally occupied by alanine.

   (b) If the tRNA was recognized by the codon, radioactive alanine would have been incorporated into hemoglobin sites normally occupied by cysteine. (The latter result was obtained by Chapeville and coworkers, who originally performed this study.)

9. (a) The formyl group of N-formylmethionine protects the amino group of methionine and prevents the derivatized amino acid from being incorporated anywhere other than at the N-terminus of the chain. In addition, N-formylmethionine is only found on initiator tRNA's, which are not recognized by EF-Tu. Thus, the formylated methionine is never bound in the ribosomal A-site.

   (b) N-formylmethionyl-tRNA$_f^{Met}$ is the only charged tRNA that binds with the 30S ribosomal subunit to form the 30S initiation complex, so methionyl-tRNA^{Met} cannot initiate protein synthesis.

10. (a) Elongation factor EF-G forms a complex with GTP that binds to the 70S ribosomal complex. Hydrolysis of GTP provides the energy needed to release the uncharged tRNA from the P-site and to cause movement of the mRNA with respect to the ribosome so that the next codon is located at the A-site. The EF-G-GDP complex then dissociates from the ribosomal complex and the next aminoacyl-tRNA can enter the A site.

    (b) The process is called translocation.

11. (a) While the basic process is the same in both systems, the initiator methionyl-tRNA$_i^{Met}$ does not require formylation in eukaryotes as is the case in many prokaryotes.

    (b) Eukaryotic mRNA molecules encode a single protein, while those of many prokaryotes encode several proteins.

    (c) Eukaryotic mRNA molecules have no Shine-Delgarno sequence for binding ribosomes, but do have 7-methylguanylate caps at the 5'-end that aid in binding the 40S ribosomal subunit.

    (d) More protein factors are involved in the eukaryotic initiation and elongation processes than are involved in the corresponding prokaryotic processes.

12. The incorporation of one amino acid residue into a protein chain consumes the equivalent of four phosphoanhydride bonds. Use the $\Delta G^{o\prime}$ value of 30. kJ/mol for phosphoanhydride bond synthesis to calculate the energy involved in the synthesis of 0.685 mole of a protein that contains 250 amino acid residues:

$$0.685 \text{ mol protein} \times 250 \text{ residues/mol protein} \times 4 \text{ phosphoanhydride bonds/residue}$$
$$\times\ 30. \text{ kJ/(mol phosphoanhydride bond)} = 2.1 \times 10^4 \text{ kJ.}$$

13. Posttranslational processing may include:

    (a) deformylation of the N-terminal residue in prokaryotic proteins,

    (b) removal of the N-terminal methionine from prokaryotic and eukaryotic proteins,

    (c) formation of disulfide bonds,

    (d) cleavage by proteinases,

    (e) phosphorylation,

    (f) addition of carbohydrate residues, and

    (g) acetylation.

D. Additional Problems

1. (a) Synthesis of the 300-residue polypeptide requires at least 900 nucleotides (NT) in the mRNA. If produced in 20 seconds, the nucleotides read per second are:

$$(300 \text{ residues} \times 3 \text{ NT/residue})/20 \text{ sec} = 45 \text{ NT/sec}$$

(b) If one protein is produced each 20 seconds by a single ribosome for each of 5000 mRNA molecules, the number of proteins produced per second would be:

$$(5000 \text{ mRNA} \times 1 \text{ protein/mRNA/ribosome})/20 \text{ sec}$$
$$= 250 \text{ proteins/ribosome/sec.}$$

If, however, the cell is producing 1000 proteins per second, the number of ribosomes reading a particular mRNA would be:

$$(1000 \text{ proteins/sec})/(250 \text{ proteins/ribosome/sec})$$
$$= 4 \text{ ribosomes.}$$

2. (a) The translation process inhibited was initiation. The Shine-Delgarno sequences in prokaryotic mRNA molecules aid in binding the 30S ribosomal subunit. Binding of the oligonucleotides by 30S ribosomal subunits prevented them from initiating protein synthesis.

(b) Eukaryotic protein synthesis was not affected because initiation of translation of eukaryotic mRNA molecules does not involve binding to Shine-Delgarno sequences.

(c) No proteins of were formed by those ribosomes bound by the oligonucleotides. Ribosomes not bound by the oligonucleotides produced normal protein chains. The overall rate of protein synthesis was decreased.

3. (a) The following table lists the codons that contain uracil and/or guanine bases, the amino acids for which they code, and the expected probability for each, expressed as a percentage. Percentages were determined as follows: For UUU, $(0.76 \times 0.76 \times 0.76) = 44\%$.

Codon	Amino Acid	Probability
UUU	Phe	44%
UUG	Leu	14%
UGU	Cys	14%
GUU	Val	14%
GGU	Gly	4%
GUG	Val	4%
UGG	Trp	4%
GGG	Gly	1%

(b) In a polypeptide of 1000 residues, there would likely be 440 phenylalanine residues, on average.

(c) In a polypeptide of 1000 residues, there would likely be 40 tryptophan residues, on average.

4. (a) Proteins destined for secretion contain an N-terminal segment called the signal peptide. The signal peptide is recognized and bound by the protein-RNA complex called a signal-recognition particle. Hence, the name.

(b) The SRP-ribosome binds to a receptor protein on the endoplasmic reticulum and the signal peptide is inserted into the membrane of the endoplasmic reticulum. This binding of the ribosome to the endoplasmic reticulum relieves the inhibition and synthesis of the protein can continue.

5. (a) In the modified residue, oxygen is replaced by selenium.

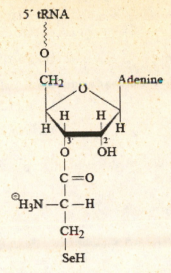

(b) The UGA codon is normally a stop codon.

Chapter 23

A. True-False

1. <u>False</u>.   They are circular, *double-stranded*, supercoiled DNA molecules.

2. <u>False</u>.   Recombinant DNA molecules are made in nature, for instance, during infection of a host cell by a bacteriophage or virus.

3. <u>True</u>.   The vector acts as the carrier molecule for the inserted DNA fragment.

4. <u>True</u>.   This is possible by the use of the enzyme reverse transcriptase. The DNA so made is referred to as cDNA (complementary DNA).

5. <u>True</u>.   Such vectors are called shuttle vectors.

6. <u>False</u>.   The complementary sticky ends will hydrogen bond with each other, but the sugar phosphate backbones of the two DNA segments must be sealed by the action of DNA ligase.

7. <u>False</u>.   The direct uptake of recombinant DNA molecules by a host cell is called genetic transformation. Recombinant DNA can also be packaged into virus particles and transferred to the host cells by the process of transfection.

8. <u>False</u>.   The transformants would be sensitive to ampicillin, but not to tetracycline. See text Section 23.2.A.

9. <u>False</u>.   The stuffer fragments are not used. The λ arms are purified and combined with the genomic DNA to create the recombinant molecules to be packaged into phage particles.

10. <u>True</u>.   The poly A tails are recognized by synthetic oligo dT molecules that associate with the poly A tail regions, by base pairing, and serve as primers for the synthesis of complementary DNA strands, catalyzed by reverse transcriptase.

B. Short Answers

1. linker

2. transformation

3. Electroporation

4. library

5. Gene therapy

6. probe

7. expression vectors

8. transgenic

9. restriction fragment length polymorphisms (RFLPs)

10. polymerase chain reaction

11. Insertional inactivation

12. transfection

13. marker gene

14. site-directed mutagenesis

C. Problems

1. The six steps outlined in your text are:

   (a) *Preparation of DNA* by purification or synthesis.

   (b) *Cleavage of DNA at particular sequences* often with a particular restriction endonuclease.

   (c) *Ligation of DNA fragments* with DNA ligase.

   (d) *Introduction of recombinant DNA into host cells* so that the recombinant DNA molecule will be propagated..

   (e) *Replication and expression of recombinant DNA in host cell* by the machinery of the host cell.

   (f) *Identification of the host cells that contain the recombinant DNA*, perhaps by identifying those cells that contain a marker gene.

2. Four types of vectors that have been used are plasmids, bacteriophages, viruses, and artificial chromosomes, such as YAC's. (One might also include cosmids, which are special types of plasmids that carry the cos sites (nucleotide sequences of the cohesive ends) of phage lambda DNA.)

3. Subjecting the solution containing the cDNA-mRNA hybrid to base-catalyzed hydrolysis would cause degradation of the mRNA chain, but would leave the cDNA strand intact.

4. RNase H would nick the mRNA strand, leaving small RNA fragments. This is a necessary step in the creation of double-stranded cDNA. (See text Section 23.5.)

5. (a) The transfection process by which phage vectors enter host cells is much more efficient than transformation.

   (b) Larger insert DNA segments can be used.

6. (a) DNA cannot be packaged into phage particles to be used for transfer to host cells if it is not of the appropriate size. To be packaged, such DNA must be about 45 to 50 kb long.

   (b) λ phage DNA must be engineered so that a sizeable piece can be removed and replaced by foreign DNA. The phage DNA is cleaved into stuffer DNA fragments and λ arms. The λ arms are used to make the recombinant DNA, but the stuffer fragments are not used.

7. DNA polymerase I has the 5'→3' exonuclease activity necessary to remove the mRNA segments that remain after use of RNase H. It also has the activity required for the synthesis of DNA to replace these RNA segments. DNA polymerase III does not have the required exonuclease activity.

8. Since cDNA libraries are constructed from mature mRNA molecules, the DNA contains only expressed genes. It does not contain introns, flanking sequences, or regulatory sequences that might be of interest to

some researchers. The cDNA libraries may, however, be more difficult to construct if they are designed to represent all mRNA molecules, including those of both high and low abundance.

9. Mature mRNA molecules have poly A tails that rRNA and tRNA do not have. The poly A tails of mRNA bind selectively to synthetic oligo dT covalently linked to an insoluble cellulose matrix. Since the rRNA and tRNA molecules do not bind to the oligo dT segments of the insoluble matrix, they remain in solution.

10. Site-directed mutagenesis of the gene that codes for the enzyme allows the alteration of a single nucleotide sequence in the gene. This leads to the replacement of a single amino acid in an enzyme by another amino acid. If the replaced amino acid is involved in the catalytic activity of the enzyme, its replacement by another amino acid residue will very likely have an effect on the catalytic activity.

11. (a) The problem of elevated temperatures is overcome by the use of a heat-stable DNA polymerase for synthesis of the new DNA strands. The DNA polymerase from *Thermus aquaticus*, *Taq* polymerase, is stable to temperatures above 90°C.

    (b) Amplification of small amounts of DNA provides enough for cloning, use as a screening probe, or sequencing. The technique has found use in the study of DNA from preserved samples (120 million-year old insects), in forensic analysis, and screening of individuals for genetic abnormalities.

12. Researchers have done this by synthesizing a short duplex DNA molecule that contains a desired restriction site. This synthetic DNA is then attached to the DNA molecule of interest (with blunt ends) by the action of T4 DNA ligase. The resulting molecule has a tailor-made restriction site that will provide sticky ends when cleaved.

13. By use of a hybrid promoter that requires a specific exogenous activator, expression of the inserted gene can be easily controlled. Bacterial cells could be cloned with the hybrid promoter turned off so that energy is not wasted by gene expression. Once sufficient bacterial colonies exist, expression of the protein product could be induced by addition of the activator. In addition, the eukaryotic protein might kill the bacterial cell, so it should be made only when cells are ready to be harvested.

14. The template mRNA molecule has a 3' poly A tail. Oligo dT fragments hydrogen bond to the poly A tail and serve as a primer for DNA synthesis toward the other end of the template molecule. Ultimately, the two strands of the duplex DNA molecule must be antiparallel, so synthesis of the DNA strand of the hybrid attached to the oligo dT primer must begin at the 5'-end and proceed toward the 3' end in order that this strand is antiparallel with the mRNA template.

15. Derived from mature mRNA, cDNA libraries do not contain intron or flanking sequences and so are much less complex than genomic libraries.

D. Additional Problems

1. Expression of the recombinant DNA molecule gave a protein product that included somatostatin attached by a methionine residue to the protein product from the *lac* gene. (AUG is the codon for methionine.) This allowed the protein product to be cleaved by BrCN to yield somatostatin itself.

2. It is easy to deduce the sequence of a DNA strand that served as a template for a particular mRNA by writing the sequence of bases that would form complementary base pairs with those of the mRNA. The sequence of the corresponding DNA coding strand is the same as that of the mRNA except for the use of thymine in locations where uracil appears in the mRNA. However, because of the degeneracy of the genetic code, several possible mRNA (and, therefore, several DNA) sequences could give rise to a particular amino acid sequence.

3. Since Trp and Met have only one codon each, but Phe, Lys, and Glu each have two codons, there are eight possible probe sequences.

4. The codons for the amino acids of the peptide are:

Trp	Phe	Lys	Glu	Met
UGG	UUU	AAA	GAA	AUG
	UUC	AAG	GAG	

Sequences for the DNA coding strand probes will be the same as those of the mRNA strands with thymine bases used instead of uracil bases. (Note: Dashes are used only for clarity.)

1. TGG–TTT–AAA–GAA–ATG

2. TGG–TTC–AAA–GAA–ATG

3. TGG–TTT–AAG–GAA–ATG

4. TGG–TTT–AAG–GAG–ATG

5. TGG–TTT–AAA–GAG–ATG

6. TGG–TTC–AAA–GAG–ATG

7. TGG–TTC–AAG–GAA–ATG

8. TGG–TTC–AAG–GAG–ATG

5. If $(10^3 \text{ bp})(100)/n \text{ bp} = 0.000035$, then $n = 10^5 \text{ bp}/(3.1 \times 10^{-5}) = 3.2 \times 10^9 \text{ bp} = 3.2$ billion base pairs.

6. Use text equation 23.1, with $P = 0.99$.

$$N = \ln(1-P)/\ln(1-n)$$

Here, n is the ratio of the size of the insert to the size of the genome, or $10 \text{ kb}/144{,}000 \text{ kb} = 6.9 \times 10^{-5}$. Therefore,

$$N = \ln(1 - 0.99)/\ln(1 - 6.9 \times 10^{-5}) = \ln(0.01)/\ln(0.999931] = 66{,}739.$$

Therefore, about 6.7×10^4 clones would need to be screened.